The Killerby, Studley and Warlaby Herds of Short Horn Cattle

The History of Their Rise and Progress

by William Carr

with an introduction by Jackson Chambers

This work contains material that was originally published in 1867.

This publication is within the Public Domain.

This edition is reprinted for educational purposes and in accordance with all applicable Federal Laws.

Introduction Copyright 2017 by Jackson Chambers

Self Reliance Books

Get more historic titles on animal and stock breeding, gardening and old fashioned skills by visiting us at:

http://selfreliancebooks.blogspot.com/

Introduction

I am pleased to present another title in the "Cattle" series.

The work is in the Public Domain and is re-printed here in accordance with Federal Laws.

As with all reprinted books of this age that are intended to perfectly reproduce the original edition, considerable pains and effort had to be undertaken to correct fading and sometimes outright damage to existing proofs of this title. At times, this task is quite monumental, requiring an almost total "rebuilding" of some pages from digital proofs of multiple copies. Despite this, imperfections still sometimes exist in the final proof and may detract from the visual appearance of the text.

I hope you enjoy reading this book as much as I enjoyed making it available to readers again.

Jackson Chambers

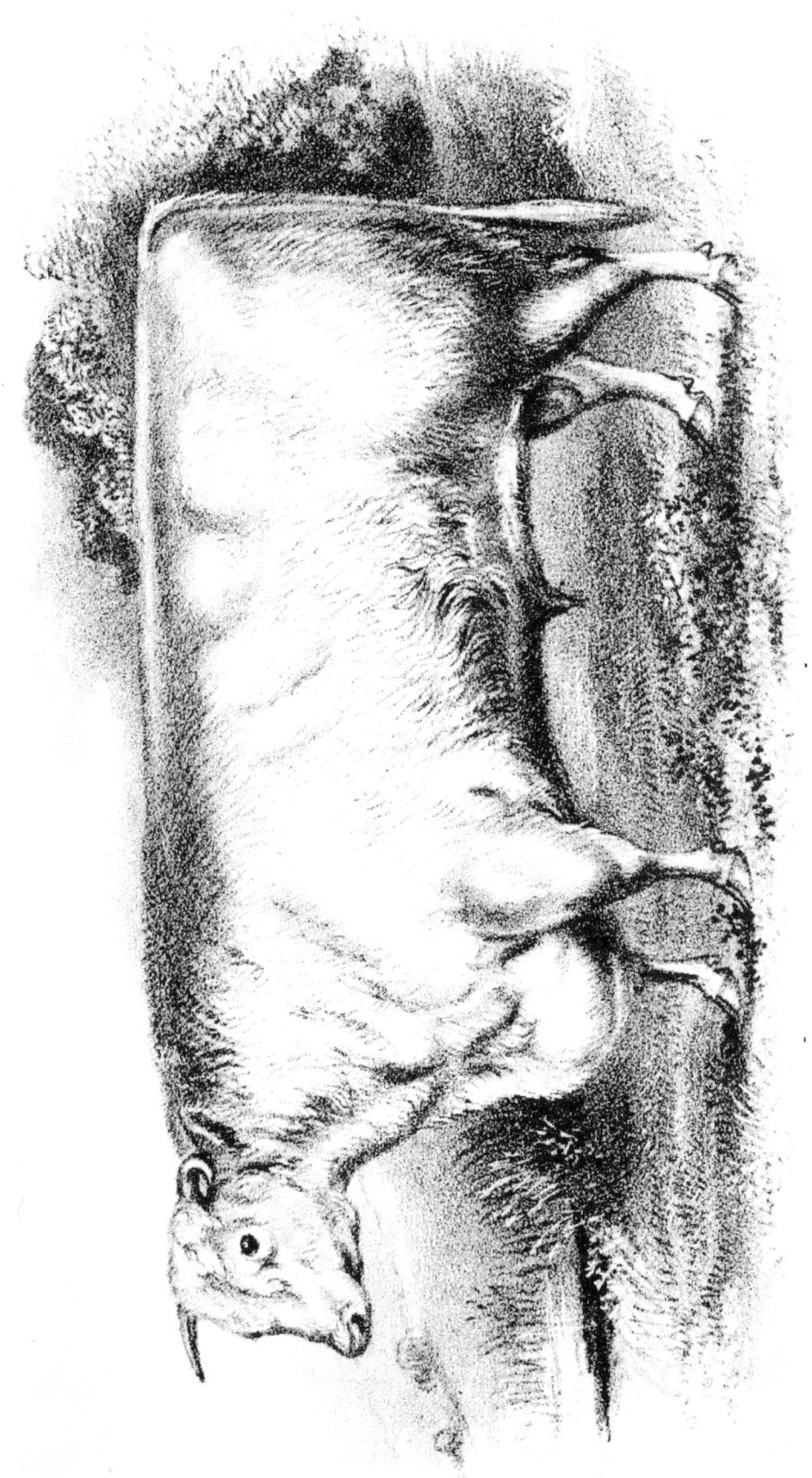

TO

RAYMOND S. BRUERE,

OF BRAITHWAITE HALL, MIDDLEHAM,

THE FOLLOWING LITTLE WORK,

SO MUCH INDEBTED TO HIS KIND ASSISTANCE,

Is Dedicated

BY HIS FRIEND,

WILLIAM CARR.

Stackhouse, Settle,
August 3rd, 1867.

PREFACE.

The history of these important herds, which I now venture to submit to the public, has for the most part already appeared in the form of letters in the pages of *The Mark Lane Express* and *Farmer's Magazine*. With the kind permission of the Proprietor of those publications, I have undertaken, at the request of numerous friends, to reprint them in a collected form. This request and the flattering reception of the letters on their first appearance, alone encourage me in the arduous attempt to do justice to a subject of such national importance as a description of the rise, progress, and present excellence of the Booth cattle, and the career, as breeders, of that well known family which has given these cattle their celebrity and their name. To the chronicle of the principal facts and events which have conduced to this celebrity, I have added, I trust in no dogmatic spirit, a few remarks on such of them as appear to illustrate the science of breeding, concluding with some general observations on the principles which seem to have guided the Messrs. Booth in effecting those gradual modifications of form and character in their cattle which have now become the distinctive and established traits of the Warlaby Shorthorn. The public, will, I am sure, make every allowance for the difficulties of a task which commences with an account of what took place at a time

PREFACE.

beyond the memory of living men, and which embraces such a very extensive field of labour. For the materials of the work, I have been largely indebted to my friend Mr. Bruere of Braithwaite Hall, whose private memoranda of events in relation to those herds had been authenticated by the late Mr. Richard Booth. I am also under great obligation to my friend the Rev. John Storer of Hellidon, for his valuable aid in the contribution of many interesting facts and suggestions, and much original matter.

W. C.

STACKHOUSE,
Aug. 3rd, 1867.

THE BOOTH HERDS.

FOREMOST among the men to whom Great Britain is indebted for the improvement of her flocks and herds, was Robert Bakewell. To him it was first given to make the grand discovery that Providence had permitted man, not only to subjugate the animal creation to his will, but also so to modify and alter the structure and conformation of the animal itself, that it might be made the more capable of ministering to his wants. The animal on which Bakewell first tried his improving hand was the sheep; and, by a careful study of Nature's laws of reproduction, and a studious selection of the best animals within his reach, he succeeded, after the labour of many years, in producing a new and improved breed, the New Leicester, the great excellence of which soon secured it universal approbation. He next carried out the same principles of improvement with regard to cattle; and what those principles were will be more fully seen in the ensuing work, for the true principles of breeding are and ever must be as immutable as the laws of nature on which they depend. Here, too, he was in a great degree successful; and if he was not so pre-eminently successful as he was with regard to sheep, and if his Improved Longhorns have not permanently established themselves as a pervading breed, it is because he had not in the Craven cattle such good material to work upon as his successors had in the Teeswater, and the Longhorns have therefore given way to the superior merit of the Improved Shorthorn.

Contemporaries of Robert Bakewell in the later period of his career, and imitators of his principles and success, were the two brothers, Charles Colling of Ketton and Robert Colling of Barmpton, both in the county of Durham, and both within a few miles of Darlington, which may be said to be the metropolis of the Improved Shorthorns, or, as they have been sometimes called, the Durhams. Of these Improved Shorthorns the brothers Colling were, about the year 1780, the most prominent cultivators. From their blood all improved shorthorns of the present day more or less derive, and many and eminent are the names of those who, imitating their success, and availing themselves of their cattle, have contributed to spread the fame of the Shorthorn through this and other lands.

With these, however, we have no immediate concern. It is my present object rather to trace the progress made in this direction by the late Mr. Thos. Booth of Killerby and Warlaby, and his sons; a history intimately interwoven with and inseparable from the history of the Shorthorn breed itself. For Mr. Thos. Booth was no servile imitator. He was a contemporary of the Collings, and began his career quite independently of them, as an improver of the cattle of the same district, and he commenced it nearly at the same time. Mr. Booth had been a breeder of shorthorns many years when the celebrated Durham ox, bred by Mr. Charles Colling, was first exhibited throughout the kingdom, and drew universal attention to the shorthorns. He afterwards did what wisdom dictated, availed himself of the Collings' best blood, and incorporated it with his own; while his sons and grandsons at Killerby, at Studley, and at Warlaby, have continued the same herd down to the present time, and given it a world-wide fame.

Previously to the year 1790 Mr. Thomas Booth, who was then the owner of the Warlaby and Killerby estates, and farmed them both, commenced at Killerby the breeding of shorthorns. Long anterior to that time a race of cattle had existed in the valley of the Tees, indigenous, it was supposed, to that and the adjacent districts, and technically known as the Teeswater. These cattle, which had long been especial favourites of the Border forayers, had excited the attention of the early improvers of horned stock, as affording favourable materials from which to make their selections. Under the auspices of several enterprising country gentlemen, foremost among whom were Mr. Milbank, Sir William St. Quintin, the Maynards, and the Hutchinsons, these Teeswater Shorthorns had undergone gradual, but not very systematic, improvement until the days of the Collings. The most favourable specimens of the breed were wide-backed, well-framed cows, deep in their fore-quarters, soft and mellow in their hair and "handling," and possessing, with average milking properties, a remarkable disposition to fatten. Their horns were rather longer than those of their descendants of the present day, and inclining upwards. The defects which Mr. Thomas Booth detected, and determined to correct in these cattle, were such as still betray the unimproved shorthorn. They were chiefly those of an undue prominence of hip and shoulder point, a want of length in the hind-quarter, of width in the floor of the chest, of fulness generally before and behind the shoulders, as well as of flesh upon the shoulder itself. The cattle had a somewhat disproportionate abdomen, a too lengthy leg, and a want of substance, indicative of delicacy, in the hide. They failed also in the essential requisite of taking on their flesh evenly and firmly over the whole frame, which frequently gave them an unlevel appearance. There was, moreover, a general want of compactness in their conformation.

Mr. Thomas Booth obtained his rudimentary stock from some of the best specimens of these Teeswater Shorthorns. He appears to have proceeded on the principle that whilst the general similitude and mingled qualities of both parents descend to the offspring, the external conformation—subject, of course, to some modification by the other parent—is *mainly* imparted by the male, and the vital and nutritive organs by the female. Acting on this hypothesis, he was careful to select such well-framed cows only as evinced, by an ample capacity of chest, a robust constitution and a predisposition to fatten, and such moderate-sized males as possessed in the highest degree then attainable the particular external points and proportions he deemed desirable to impress upon his herd. A dairy farmer under Lord Harewood, a Mr. Broader, of Fairholme, in the parish of Ainderby, appears to have possessed some cows having the qualifications required. Tradition speaks of them as unusually fine cattle for that period; good dairy cows, and great grazers when dry; somewhat incompact in frame, and steerish in appearance, but of very robust constitution.

Previously to the year 1790, Mr. Thomas Booth had bought some calves from these cows. Strawberry Fairholme, Hazel (*i. e.* flecked roan) Fairholme, and Eight-and-twenty-shilling Fairholme, purchased from Mr. Broader's farm, have the honour of being the ancestresses of several illustrious families of shorthorns.

I have said that Mr. Thomas Booth selected *moderate-sized* males. His observant eye had recognised, as indispensable to any improvement in the symmetry of these Teeswater animals, the necessity of reducing in size and stature their large, loosely-knit frames. With this view he decided on selecting his bulls from the stock of his contemporaries, Messrs. Robert and Charles Colling, who had themselves, to some extent, effected this reduction of size, and improvement of form and fattening capacity in their stock, chiefly through the use of Hubback, a small short-legged bull. Twin Brother to Ben (660), bred by the Collings, and Son of Twin Brother to Ben, were the first bulls used by Mr. Thomas Booth to these Fairholme heifers. These bulls had the short legs, the long and level hind quarters, the firm backs and good twists, to which Mr. Thomas Booth attached so much importance, and their offspring amply testified to his discrimination. It is recorded that one cow by the former, and her daughter by the latter bull, produced six calves in one year, the dam having twice produced twins, and the daughter once. Four of these calves were heifers. Some of the offspring were very superior cows. In proof of the excellent foundation they afforded for the formation of a herd, it is affirmed on high authority that one of the Twin-Brother-to-Ben cows produced, to Son of Twin

Brother to Ben, a cow quite equal to Faith by Raspberry, the dam of the famous Hope. Many of the cows were deep milkers, but running dry sooner than was then usual, when they gained flesh very rapidly. The late Mr. Ewbank, of Sober Hill, questioning the milking capacity of some of them in this condition, Mr. Thomas Booth pointed to their broad backs, and exclaimed, "Look there! that is worth a few pints of milk!" These cows were further open to Mr. Ewbank's criticism as having *raw* noses, as he contemptuously termed that feature when flesh-coloured; alleging that in *his* early days the farm stock was nearly all *black-nosed*, and that he never knew a raw-nosed cow that was not delicate,— a prejudice which has long since passed away.

Having thus judiciously selected the best animals procurable of both sexes, Mr. Thomas Booth was careful to pair such, and such only, of the produce of these unions as presented in a satisfactory degree the desired characteristics, with animals possessing them in equal or greater measure, and unsparingly to reject—especially from his male stock—all such as were not up to the required standard. Having by these means succeeded in developing and establishing in his herd a definite and uniform character, he sought to ensure its perpetuation by breeding from rather close affinities, as in his opinion the only security for the unfailing transmission, and transmission in an increased ratio, of these acquired distinctions to the offspring. In tracing the pedigress of these herds, it will be seen that from the earliest period the same system of breeding from close relations which was pursued by the Collings was followed by the Booths.

An examination of the pedigree of Lady Maynard (*alias* the cow Favourite) will show to what a length the system was carried by the early breeders, and how closely the first families of the Colling strain were allied to the Booth tribes. Further proof of this may be found in the pedigrees of the earliest bulls used by Mr. Thomas Booth, namely, Twin Brother to Ben, Suworrow, Albion, Pilot, and Marshal Beresford. Take, for example, the three last named. Albion—purchased at Mr. Charles Colling's sale in 1810, by Mr. T. Booth, senior, for 60 guineas, when a calf—was by a bull which was both a son and grandson of Favourite; his dam was by a son of Favourite, and his grandam by a bull who was not only a son of Favourite but also of Favourite's half-sister. Pilot, bred by Mr. Robert Colling, was by Major or Wellington. Major was by a son and grandson of Favourite, his dam by a son of Favourite, his grandam by Favourite, and his great-grandam by Favourite. Wellington was by a son and grandson of Favourite, and his dam was by Favourite. Marshal Beresford was by a son and grandson of Favourite, his dam by a grandson of Favourite, and his grandam by Favourite.* Marshal Beresford came into the herd in an exchange for some cows with Major Bower, Mr. Thomas Booth's brother-in-law, a shorthorn breeder, then living at Welham. On returning home one day, Mr. R. Booth found, to his great annoyance, that his father had resold the Marshal to Major Bower. He thought that if either had been

---

* On this subject the reader is referred, for a particular consideration of this point, to a very excellent and interesting communication from the Rev. John Storer. See Appendix.

parted with it should have been Albion. It proved fortunate, however, for the Booth herd that Albion was retained; for though not so stylish as the Marshal in appearance, he proved far superior to him as a sire. Albion is said to have done more good in the herd than any other of the earlier bulls, notwithstanding that he had, through Washington (674), *the alloy*, which was the term of reproach cast upon Lady by Grandson of Bolingbroke and her descendants in the early days of shorthorn breeding. The offspring of Albion were, in general, very round, compact, and near the ground.

I must here, however, revert to the Fairholme calves. A slight survey of the tribes which have sprung from these early mothers of the herd may not be without interest to some of my readers. From them proceeded the Fairholme or Blossom tribe, the old Red Rose tribe, and the Ariadne or Bright Eyes tribe.

Of the Fairholme or Blossom tribe, one branch terminated in the bull Easby (232). Another, which Mr. R. Booth took with him to Studley, produced Moss Rose by Suworrow, Madame by Marshal Beresford, Fair Maid by Pilot, Miss Foote by Agamemnon, and Young Sir Alexander. A third division, which, in the cow Eve passed into the hands of Major Bower, has representatives in the herd of Lord Feversham—Skyrocket, the first prize bull at the Royal show at Leeds in 1861, being one of them. Of a fourth branch—the descendants of Beauty by Albion—one portion remained in the hands of Mr. John Booth, and produced Modish, sold to Mr. R. Holmes (who bred from her Belzoni); the other passed into the hands of Sir Charles Knightley, who had at one time several representatives of it. From a fifth branch, retained by Mr. Thomas Booth, sprang Twin Cow by Albion, her son Navigator, whose spirited portrait adorns the walls of the dining-room at Warlaby, and a long array of prize animals, amongst which may be mentioned Bloom, Plum Blossom, Nectarine Blossom, Venus Victrix, Baron Warlaby, and Windsor.

The old Red Rose tribe is extinct, except in the progeny of Julius Cæsar and Belshazzar.

From the Bright Eyes tribe, in the possession of Mr. R. Booth, at Studley, came Ariadne, the prize cow Anna by Pilot, and many other fine animals dispersed at the Studley sale.

Besides these Fairholme tribes, there was the Halnaby or Strawberry tribe, which also dates from this period. The first of them was of that yellow red and white hue, which, though out of favour at the present day, was then the prevailing colour of the shorthorn. She was bought in Darlington market, and one of the earliest recollections of Mr. R. Booth was of that cow coming home. The type of old Halnaby of 1797, who is said to have been a very finely-made cow, has often been reproduced in her descendants in the herd. Mr. Thomas Booth considered this as one of his finest families, quite equal to the Blossom and the Ariadne tribes. Young Albion (15) is the first bull of note in the Halnaby family. He was much used in the herd, and was one of the first that was let out on hire. He went to Mr. Scroope's, of Danby Hall, near Middleham, who had a fine, large, robust herd of cattle, related, through some of the bulls used, to the Colling blood. In 1812, the Squire of Danby challenged

Mr. Thomas Booth to show, "for rump and dozen" (the usual stakes at that day being rump steaks and a dozen of wine) the best lot of heifers he had, against the same number of his own, the match to be decided at Bedale. Although a good lot, the Danby had to give place to the Killerby and Warlaby contingent. Of the Halnaby tribe came also the bull Rockingham, and Priam, the sire of Necklace and Bracelet. The only female representatives of the family are in the hands of the present Mr. Booth of Warlaby. From Strawberry 3rd came the Bianca and Bride Elect branch; whilst the famous cow White Strawberry, the dam of Leonard, was the ancestress of Monk, Medora, Red Rose, and her daughters, the queenly quartette. Young Matchem (4422) is descended from White Rose, own sister of Young Albion, and therefore, on the dam's side, of the Halnaby family, and the same branch of it gives the dam, Young Rachel, of Mr Ambler's Grand Turk.

The Bracelet tribe sprung from a cow by Suworrow, of whose origin there is no record. She was the ancestress of a very superior cow, calved in 1812, Countess by Albion, the alloy bull; also of Toy, and her twin daughters Necklace and Bracelet, and of Col. Towneley's Pearly, and Mr. Torr's Young Bracelet tribe.

The early representatives of the above-mentioned tribes formed the herd of Mr. Thomas Booth down to the year 1814, when Mr. Richard Booth, taking the Studley Farm, near Ripon, left Killerby. Mr. Thomas Booth was at this time the most enterprising and skilful improver of cattle in his district, if not of his day. It is said there were some cows in Mr. Thomas Booth's herd of that period as good as any herd of the present time can boast; though, being bred for use rather than show, the generality of them were wanting in the refinement of the modern shorthorn. At that period there were, happily, no shows to demand the sacrifice of the best cattle in the kingdom, or the few that were held could be reached by the majority of cattle attending them only by such long journeys on foot as would be impracticable by animals in such a state of obesity as is now a *sine qua non* with the judicial triumvirate. High feeding at that time meant no more than good pasture for cows early dried of their milk; and the term "training" was never heard except in relation to horses. The first breeder who introduced the system, which has since run into such ruinous excess, of house-feeding cows and heifers in summer on artificial food, was Mr. Crofton; and in that year he, of course, took all before him in the show yards. The general treatment of the females of a herd at that day was a simple hay diet during the winter months. They were put early to breeding, and generally calved at two years old. A few were taken from the lot to milk. The remainder suckled their calves until winter. They were then taken up, dried, and fed off by the time they were three years old; the same course being pursued, in their turn, with their progeny.

. Mr. Thomas Booth was as liberal as his successors in allowing the free use of his bulls to his poor neighbours; and, like most public benefactors, was occasionally imposed upon. A ludicrous instance of this is still remembered. An old fellow at Ainderby, not contented with the bull set apart for this purpose, and being anxious to have a calf by another, that Mr. Booth especially prized and kept exclusively for his own herd, took his cow into the lane adjoining the field where the prohibited animal was grazing. The bull broke through the fence; and—the old Yorkshireman's object was achieved. The latter, knowing how indignant Mr. Booth would be, thought it safest to act on the principle of taking the bull by the horns; and, assuming an injured air, at once repaired to him, exclaiming, "O maister, maister! sic an a thing has happened! Your gurt ugly beast has broken through t'hedge, and I doubt he'll hae gitten my cow wi' cauf. It's a sad bad job; for I were boun' to feed her off."

Mr. Richard Booth's removal to Studley forms a new era in the history of these herds. From 1814 down to its dispersion in 1834, the Studley colony took precedence of the parent stock. We may now, therefore, before proceeding with the history of the Killerby Herd, turn our attention to that of Studley.

THE STUDLEY HERD.

Mr. Richard Booth inherited with his father's name his full share of his father's skill as a Breeder, with an equal fondness for the pursuit; and his new farm, which he held under the wealthy and well-known Mrs. Lawrence, was speedily stocked with superior shorthorns. He began with his father's cattle, and carried on to even greater perfection his father's work. Among the first importations which were made from Killerby to

Studley, when Mr. Richard Booth went there in 1814, the following may be mentioned:—He purchased from his father Bright Eyes, by Lame Bull, and her daughters Ariadne, then a two-year-old, and Agnes, a yearling, both by Albion. Ariadne was own sister to Agamemnon, the grandsire of Isabella by Pilot. She was the dam of the famous Anna by Pilot, who won numerous prizes at the best shows of the day; and who, in 1824, performed the feat of walking from Studley to Manchester, taking the first prize there, walking back, and producing within a fortnight Young Anna. Anna is said, by those who well remember her, to have borne a very strong resemblance in colour and character to Queen of the Ocean. She was the dam of Adelaide, who, through her sire Albert, was also grandaughter of Isabella. Adelaide was the highest priced female sold at Mr. R. Booth's Studley sale in 1834, and was the grandam of Mr. Storer's cow Princess Julia. From Anna, more remotely through her daughter Young Anna, are descended two of Mr. Torr's families; and from Agnes, daughter of Bright Eyes, came Mr. Fawkes's Verbena and her descendants. Agamemnon, the own brother of Ariadne, was a bull of extraordinary substance, with good hind quarters, heavy flanks, deep twist, and well covered hips. He was eventually sold, with two heifers, to Mr. White, of Woodlands, near Dublin. Even in these early days Mr. Booth had bulls out on hire. Alonzo, a son of Ariadne, by Rockingham, was let to Mr. Hutton, of Marske, who, to promote the improvement of the breed of cattle in his district, had at that time yearly shows on his estate. Protector, another bull of the Bright

Eyes family, was hired by Mr. Powlett, of Bolton Hall. He was a large red bull, and a capital sire.

In the first year of his residence at Studley, Mr. R. Booth bought in Darlington market the first of what was afterwards known as the Isabella tribe. She was a roan cow, by Mr. Burrell's bull of Burdon, and, for a market cow, had a remarkably ample development of the fore-quarters. She was put to Agamemnon. The offspring was "White Cow," which, crossed by Pilot, produced the matchless Isabella, so long remembered in show-field annals, and to this day quoted as a perfect specimen of her race. Pedestrians crossing the fields to the ruins of Fountains Abbey might generally see her and Anna, perhaps the two best cows of their day, with a blooming bevy of fair heifers, attended by Young Albion; and many a traveller lingered on his way to admire their buxom forms, picturing to himself perhaps how the monks of the old abbey would have gloried in such beeves. Isabella was the Rev. Henry Berry's beau ideal of a shorthorn. In 1823, Sir Charles Morgan having offered a premium to promote a trial of merit between Herefords and Shorthorns, Mr. Berry wrote to the editor of the Farmers' Journal requesting him to give publicity to the following offer: "I will produce as a competitor for Sir Charles Morgan's premium at Christmas next a Shorthorned cow, then nine years old, expecting to drop her *eighth living calf, at separate births*, in June now next ensuing, against any Hereford in England, seven or nine years old, having had calves for years in the same proportion. I will also on the same occasion produce a Shorthorn heifer, three years old, having had a living calf, allowing to the Herefords the same ample scope—*all England*—for the production of a competitor. It will be obvious to your readers that in thus pitting two individuals against so numerous a tribe as the Herefords, I must entertain considerable confidence in their merits, and it will be as easy to draw a correct conclusion should my offer not be accepted." The cow and heifer which, by permission of the owners, Mr. Berry proposed bringing into competition with the Herefords, were Mr. Whitaker's cow Moss Rose and Mr. R. Booth's heifer Isabella by Pilot. The challenge was not taken up.

Isabella and her descendants brought the massive yet exquisitely moulded fore-quarters into the herd, and also that straight under-line of the belly, for which the Warlaby animals are remarkable. That such a cow should have had but three crosses of blood is striking evidence of the impressive efficacy of these early bulls, and confirms Mr. R. Booth's opinion that four crosses of really first-rate bulls of sterling blood upon a good market cow, of the ordinary Shorthorn breed, should suffice for the production of an animal with all the characteristics of the high-caste Shorthorn. In such an opinion, confirmed by such an example as this, there is much instruction and encouragement for tenant farmers desirous of improving their stock. Female Shorthorns of high pedigree are in general beyond the reach of their class; but if near neighbours would but club together, and procure for their joint use a succession of pure bred males, of fixed and determinate character, the improvement in a few years effected in their stock, especially as regards early maturity and tendency to carry flesh, would be such as materially to enhance their farming profits. It is often difficult, however, to convince even those amongst them who have used, and experienced the benefits resulting from the use of a high bred sire, of the expediency of *continuing* in the same course. Some wretched cross-bred cow put to the "pedigree bull," probably produces a bull-calf with all the characteristics of its sire, all the more probably, perhaps, from her being of no distinctive character herself. This the farmer rears on something better than blue milk, in the hope of getting a prize or two with him at local shows, nine out of ten of which absurdly ignore the first desideratum in a sire, pure descent, the bull of one cross being allowed to compete with the possessor of half-a-dozen. The mongrel gets the white ribbon, and immediately becomes, in his owner's estimation, endowed with every necessary qualification for a sire. The farmer thenceforth uses him to his own cattle, and perhaps those of half the neighbourhood. The result of this retrograde step is soon apparent in the stock. Interesting traits of the maternal ancestry of the *parvenu* bull re-appear in his progeny—the brindle, it may be, of Pat O'Flanaghan's Kerry, the black nose and horns of Sandy Macpherson's Kyloe, or the long legs and flat sides of Taffy Owen's Glamorgan. But though the farmer sees that he is rapidly losing all the ground he had gained, and that his stock has ceased to be sought after, he rarely admits the cause.

"White cow," by Agamemnon, produced, besides the famous Isabella, "own sister to Isabella," and Lady

Sarah, and was then sold to Mr. Paley, of Gledhow. Her dam, the Darlington cow, had previously been disposed of to the master of a boarding-school at Ripon, one of whose pupils, Mr. Bruere, of Braithwaite Hall—a highly esteemed friend of the late Mr. Booth's—well remembers the brimming pails of milk she gave. "Own sister to Isabella" was the dam of Blossom, by Memnon (a son of Julius Cæsar and Strawberry by Pilot), and Blossom was the dam of Medora by Ambo, one of the neatest cows Mr. Booth ever bred. Medora was sold to Mr. Fawkes, in whose hands she was the progenitress of his Gulnare, Haidee, Zuleika, and others. Mr. Fawkes's Lord Marquis, the first prize three-year-old bull at the Royal Show at Lewes in 1852, and the Yorkshire Show at Sheffield in the same year, was also a descendant of Medora's. "A gentleman," says the writer of 'Shorthorn Intelligence,' "who has been intimately conversant with the herds of Great Britain for at least a quarter of a century, declares that one of the most interesting sights he ever saw at an Agricultural Exhibition was on the show-ground at Otley some years ago, when, after the judging, the famous Booth cow Medora by Ambo was led round the ring, followed by her six daughters, all of them, as well as the mother, decorated with prize favours. The daughters were Gulnare, Haidee, and Zuleika (by Norfolk); Victoria, and Fair Maid of Athens (by Sir Thomas Fairfax); and a heifer named Myrrha, not in the herd book, under that name at least, by Rockingham (2550)." Blossom was bought by the Earl of Lonsdale at the Studley sale in 1834, and, after breeding four calves, was slaughtered in 1840.

Own Sister to Isabella also had Imogen by Argus, which was sold at the Studley sale to the Earl of Carlisle, and became the dam of Isabel, by Belshazzar (1704). This Belshazzar, who was contemporary with Mr. Booth's Belshazzar of the old Red Rose tribe, was from Lady Sarah, the third sister of Isabella by Pilot. Lady Sarah became the property of the Earl of Carlisle, and produced at Castle Howard three bulls and four heifers, one of which was the dam of Lord Stanley, purchased by Messrs. Booth and Maynard.

Isabella by Pilot, now the best known to fame of the three sisters, produced, at Studley, Isaac by Young Albion, Albert by the same bull, Isabella, sold to Mr. Bolden, Young Isabella to Mr. Paley, and Belinda to the Earl of Carlisle, and four others; and on the sale of the Studley herd she alone was retained, and transferred to Warlaby, where she gave birth, in her eighteenth year, to Isabella Matchem, afterwards the dam, as will be seen, of a numerous progeny. The fate of Isabella's son Isaac was always looked back to by his breeder with regret. The demand for bulls was then only commencing. Isaac had been let for a year to Miss Strickland, of Apperley Court, and on his return, Mr. Booth not requiring him, he was unfortunately fed off to make room for younger ones, before his eminent merits as a sire had been discovered. The Isabellas had all great capacity for rapidly acquiring ripe condition on pasture. As an illustration of the fallaciousness of the usual mode of judging cattle by the softness of their flesh, it may be worthy of mention that at one of the Yorkshire Agricultural Meetings held at Northallerton, a grass-fed heifer, a daughter of Isabella, by Ambo, was shown, and rejected as being too hard-fleshed. Not breeding, she was slaughtered at York for Christmas beef. Her two successful rivals also failing to breed were slaughtered, and the palm for the best carcase of beef was awarded to Mr. Booth's heifer over her Northallerton rivals. Nor is this case without many a parallel in the history of Royal Shows. Numerous as have been the prizes which the Booth cattle have received, their number would have been greatly increased if judges had always carefully distinguished between flesh and fat. When their decisions have been on this ground—as they often have been—adverse to the Booth cattle, many an experienced butcher has proclaimed a very different opinion; and could the appeal *ad crumenam* have been adopted by an immediate sale of the rival animals to the shambles, how useless would it have been in most instances to contest the supremacy of the Booths!

Another cow which Mr. Booth took with him to Studley was Madame, by Marshal Beresford, also of the Fairholme Blossom tribe. From her came Fancy and Fair Maid, both by Agamemnon. The former was the dam of Fatima, a very neat, middle-sized cow, which, put to Mr. Maynard's Sir Alexander, produced the famous bull Young Sir Alexander. This bull was the sire of Strawberry, whose daughter White Strawberry by Rockingham held perhaps equal rank in Mr. Booth's estimation with Anna, Isabella, and her own contemporary rivals, Necklace and Bracelet. Fair Maid, the other daughter of Madame by Marshal Beresford, was the dam of Miss Foote, whose descendants were very numerous, and were all disposed of at, or previously to, the Studley sale. They united in a remarkable degree the two properties of good milking and rapid fattening. Fair Maid herself was sold to Mr. Ellison of Sizergh, where she bred many calves, and proved herself an excellent dairy cow. Miss Foote was sold to Captain Shawe, and Fair Helen, her daughter, who was the dam of the noted bull Cossack, to Sir Charles Tempest, with whom she bred four heifers. I remember, in 1853, a stray waif of this famous tribe in the hands of an innkeeper, at Clapham, in Yorkshire. It was, in fact, the broad level back and symmetrical proportions of this cow that induced me to purchase my first shorthorn, her bull calf. The cow was a grandaughter of Miss Foote, being a daughter of Lady Helen, then the property of Mr. Foster of Clapham. She was sacrificed whilst still in her prime, her owner being tempted by the offer of a high price for her from a butcher.

Some mention of the bulls bred and used by Mr. Booth during his residence at Studley seems here to be required.

One of the first bulls of superior mark bred by Mr. Richard Booth after his removal to Studley was Julius Cæsar, a bull of very symmetrical proportions, which he had the merit of impressing in a surprising degree upon his offspring. No matter how dissimilar and opposite in form and breed the cows to which he was put might be, the produce all bore the unmistakeable stamp of their sire. The offspring, by him, of the shabbiest lane-side cow, had, it is said, all the character of the pure-bred Shorthorn. It may be worth while to inquire how far the remarkable property which distinguished this bull may be traced to the preponderating influence of any particular progenitor or progenitors in his pedigree, an investigation of which, it may here be sufficient to say, will show him to be descended half a dozen times, and some of them very nearly, from Twin Brother to Ben.

This circumstance lends weight to the opinion of many experienced breeders, that, in general, the capability of a bull to transmit to his offspring his own peculiar mould and properties depends upon his having inherited them from a succession of ancestors endowed with similar characteristics. It is doubtless to the concentration of hereditary force thus derived that the extraordinary transmissive power of such bulls as Comet, Favourite, and Julius Cæsar is to be attributed. At the same time it is a curious circumstance, and one that should not be forgotten—as often modifying to some extent the principle above enunciated—that amongst animals similarly bred there are some bulls, and some cows too, that possess an immeasurably greater transmissive influence than pertains to others.

Pilot, another of the bulls of this period, was bred by Mr. R. Colling, and purchased by Mr. T. Booth at the Barmpton sale in 1818 for 270 gs. He was used in all the three herds, and there was no bull to which they were more largely indebted. The close in-and-in breeding of this animal has already been shown. He was let to Mr. Rennie for a short time; but his stock at home proved so good, that he was recalled at the expiration of his first season. Pilot was a small compact bull, somewhat undersized, but possessed of great thriving propensity. He was a capital sire, and may be appropriately cited as a striking example of the preceding remarks. I am indebted for this account of Pilot to one who remembers him well—that old friend of the Booths, the much respected Nestor of the Shorthorns, Mr. Wetherell, who, like his friend Mr. Wiley of Brandsby, is still hale and strong, a living record of early Shorthorn times, from whom younger men learn the lessons of the past. Isaac, another bull of note, bred by Mr. Richard Booth, has already been referred to. Burley and Ambo, or Ambo Dexter, both containing a large amount of the Favourite blood, were partially used in the herd during the last three years before the sale.

In the year 1834 Mr. Richard Booth, finding that some of his best pastures were required by their owner for other purposes, gave up the farm at Studley, and selling off the whole of his herd, with the exception of Isabella by Pilot, retired to Sharrow, near Ripon. After residing there for a year, which, from being bereft of his favourites, he used to describe as the least happy period of his life, Mr. R. Booth, in consequence of his father's death, succeeded to the estate and shorthorn herd of Warlaby. The sale of the Studley herd was a step which Mr. Booth always regretted, for many of the animals it contained were, in his opinion, every whit as good as any he afterwards bred. They were dispersed into many hands, and though Old Cuddy's assertion, that they have " a' swealed away," is certainly too sweeping, it may be doubted whether, even in the hands of very celebrated breeders, like Mr. Fawkes and others, the descendants of these famous cattle have ever quite equalled their cousins at Warlaby.

It is now necessary to go back a quarter of a century to resume the history of the Killerby Herd.

THE KILLERBY HERD.

We have seen that in the year 1814 Mr. Richard Booth took with him to Studley some of the animals then forming the Killerby herd. Mr. Thomas Booth shortly afterwards supplied the place of these with other cows, which became the foundresses of three famous tribes—the Farewell tribe, from which sprang Faith, Hope, and Charity; the Broughton tribe, from whence came Bliss, Blithe, and Bonnet; and the Dairymaid, or Moss Rose tribe, from which are descended Vivandiere, Camp Follower, and Soldier's Bride. The first of the Farewell tribe came from Darlington; the first of the Broughton tribe from a dairy farmer in a village of that name, who had some good cattle, but, pedigrees being slightly valued in those days by the tenant-farmer class, nothing further is known about them. The first of the Dairymaid tribe came from an equally good stock in the village of Scorton.

In the year 1819, on the occasion of Mr. J. Booth's marriage, Mr. T. Booth removed to Warlaby, giving up to his son, Mr. J. Booth, the Killerby estate and a part of the shorthorn herd, and taking the remainder with him. A portion of the Fairholme or Blossom tribe, and of the Old Red Rose tribe, were removed to Warlaby, the remainder being left with Mr. John Booth. The Halnaby family was also divided, but the famous Bracelet tribe was all left at Killerby. From this period down to the year 1835, when Mr. R. Booth succeeded to his father's herd at Warlaby, there is comparatively little known of the two herds. The times were unpropitious for the Shorthorn. The spirit of improvement which the example of the Collings had evoked only partially survived. There was a general depression in all agricultural produce, and consequently but little demand for animals, the purchase of which appeared at that time to partake so much of the nature of a speculation. Not yet did

"Generous Britons venerate the plough,"

or regard with respect bucolic occupations. A man gained more *eclat* by a display of science and judgment in going across country than in the breeding of cattle. In some districts, a gentleman almost lost caste by devoting himself to such ignoble pursuits, and was sarcastically dubbed, by his companions in the pink, "cow-scratcher."

But though "fallen on evil days," the stock at Killerby was of high character, and was frequently resorted to by the few good breeders of that period for the purchase of animals. It is a house where all comers were, and still are, regaled with the welcome of the olden times. Killerby is one of the pleasantest of the pleasant homes of England. It is a substantial square manor house, picturesquely situated on a gentle eminence to the south of the river Swale, and two miles from Catterick, the site of the once important Roman camp and city of Cataractonium. The house occupies the site of the ancient castle of Killerby, once a stronghold of great magnitude, founded in the reign of Edward the First by Sir Brian Fitzalan, Earl of Arundel. It is approached by a road winding through verdant pastures thrown together into the form of a park, adorned here and there with noble elm and walnut trees. The estate consists of about 500 acres of arable and pasture land. The soil, which is very mixed—gravel, strong clay, marl, and peat being sometimes found in the same field—is more adapted for sheep than heavy cattle, though there are two or three excellent pastures. Several of the inferior grass fields have been ploughed up of late, and heavy crops of oats and turnips grown in their place, which has allowed the number of sheep kept to be greatly increased. Although half-bred sheep are occasionally seen on the farm for summer grazing, the staple stock are pure Leicesters, for the wool of which Mr. J. B. Booth, the present owner, has gained several prizes at the Yorkshire shows.

The late Mr. Booth, of Killerby, was known and beloved throughout the county as a strikingly genial example of the worthy and hospitable northern agriculturist, ever devoting himself to the service of his friends (and he had many) and to the advancement of agricultural improvement. The humblest, equally with the most important, agricultural societies might always rely on his good offices, whether as patron or judge, in which latter capacity being confessedly unrivalled, he was in great request, and would most good naturedly consent to officiate, though his doing so involved the exclusion of his own cattle from competition. As might have been expected, from his fine and manly character, he was also a keen sportsman; like Chaucer's squire,

"Well could he sitte a horse and faire y-ride;"

and Yorkshire, that modern Thessaly of horsemen, knew no more thorough judge of hack or hunter. His skill in this respect still survives in his sons; many a field and many a showyard testify that in this regard, as in others, Killerby has not degenerated from its ancient fame. He had, too, a natural taste for the fine arts, and when from illness he could not go far from home, he had his horses led out, and would sit on the lawn, or in the hall, to paint them. Here, too, his taste survives, and if I touch lightly on the subject it is because more delicate fingers now hold the brush, and I would not trespass unbidden upon the elegant recreations of Killerby's fair Mistress.

When, on the establishment of the national shows in 1839, the superiority of the Killerby Shorthorns had been proved in contest with the best animals of the day, the herd attracted many visitors, and its inspection was as free to all classes as were the fruits of its owner's experience in breeding, which he was ever ready to communicate to the neophyte. It may not be uninteresting to the present fair enthusiasts in shorthorn matters to learn, that in the absence of her husband, the late Mrs. Booth—a lady who will long be remembered in that neighbourhood for her benevolent disposition and engaging manners—would herself most affably do the honours of the herd, leading the way to her especial favourites, and expatiating on their pedigrees, points, and perfections, sometimes with a dash of arch humour, and always with the grace and delicacy of the thorough-bred lady that she was. Mrs. Booth's sister, Miss Wright, had an equally keen appreciation of the merits of a good shorthorn, and would stop any one of kindred tastes, who happened to be passing through Cleasby, to have a chat on her favourite topic, or to lead them to the Garth (since known by his name), where in the fulness of his days and honours repose the remains of Comet.

For some years after the dispersion of the Studley Shorthorns, in 1834, (until perhaps 1845-6, when Mr. R. Booth's Faith and Hope were in the ascendant), the Killerby herd, with the halo of its Bracelet and Mantalini triumphs, held a more conspicuous place in public view than that of Warlaby, and it is fitting, therefore, that we should first turn our attention to the history of its tribes, though the two herds, owing to their common origin and the constant interchange of sires, are so intimately allied as to make it difficult to dissociate one from the other: thus, for example, Pilot was joint property and used in both herds; Mr. R. Booth's Argus and Priam were respectively the sires of Toy and her daughters Necklace and Bracelet in the Killerby herd, while Bracelet's son Buckingham became the sire of Charity and a host of prize winners in that of Warlaby. Again; Mr. R. Booth's Leonard, Vanguard, and Hopewell were freely used in both herds.

Amongst the cows composing the Killerby herd in the year 1835 were Toy by Argus, her daughter Little Toy by Volunteer, and Ivory by Matchem, all of the Bracelet tribe, and three other Matchem cows, namely, Maiden of the Mantalini tribe, Landlady of the Lady Betty tribe, and Floranthe the grandam of Colonel Towneley's Beauty.

Toy was a very neat, thick-framed cow, with a magnificent udder. Her milking capabilities were the boast of the Killerby dairymaids, and were transmitted to her famous twins Necklace and Bracelet. The latter were bred from very close affinities; thus Vestal, the dam of Toy, was by Pilot; Argus, the sire of Toy, was by Young Albion out of Anna by Pilot; Toy's two daughters,

Necklace and Bracelet, were by Priam, a grandson of Pilot on his dam's side, and whose sire, Isaac, was by young Albion, from Isabella by Pilot. Toy had twice previously been put to Young Matchem, but the offspring, Teetotum and Plaything, were not at all equal to the twins. Teetotum, however, was the parent of Ladythorn by Lord Stanley, whose portrait in the "Herd Book"—evidently a portrait—unmistakeably proclaims her a good cow; and in truth she was second only to Birthday in merit. She was sold to Mr. Banks Stanhope for 150 guineas, a great price in that day, and in 1845 won the first prize at the Royal Exhibition at Shrewsbury in the cow class; her two daughters, Ladybird and Rovesby Thorn, subsequently earning fame in the show-fields. From Toy was also descended the cow Gertrude, purchased, together with her lovely daughter Lady Hopetown, at Mr. Bolden's sale, by Mr. Torr, who had previously bought, at the Killerby sale, Sylphide by Morning Star, also sprung from Toy. Some others of her descendants are in the hands of Mr. Pawlett, and Mr. Torr's noble bull Breastplate has well kept up the honour of the family. But I must here return to the twin-born progeny of Toy, the all conquering Necklace and Bracelet.

It does not appear that Mr. John Booth was a very frequent competitor in the show-field until the establishment of the Royal and Yorkshire Shows in 1839. Before this time shorthorn cattle were kept chiefly for dairy and grazing purposes; the majority of the male stock were steered, and many a fine heifer that took the butcher's eye was converted into Christmas beef. Necklace and Bracelet shared the pasture and the straw-yard with the ordinary stock of the farm until nearly two years old. As calves they never had more milk than their dam, who suckled them both, supplied; and, throughout the whole of their victorious career, they derived their chief support from the pasture, with a daily *bonne bouche* of corn and cake. Yet Bracelet won seventeen prizes at the various meetings of the Royal Agricultural Society of England, the Highland Society of Scotland, the Yorkshire Society, and other local shows; and at the Yorkshire Show in 1841, where she won the first prize for extra stock, the sweepstakes for the best lot of cattle not less than four in number, was awarded to Bracelet, Necklace, Mantalini, and Ladythorn. Necklace won sixteen prizes and one gold and three silver medals at the various meetings above mentioned, as well as at the Smithfield Club, where she finished her career as a prize-taker in 1846 by winning the first prize of her class and the gold medal (for which there were thirty-seven competitors) as the best animal exhibited in any of the cow or heifer classes. At the Smithfield Show, in the following year, the same prizes were awarded to Mr. Wiley, and in 1849 to Mr. Cartwright, for animals bred from the Killerby stock. In five years four first prizes for the best Shorthorn cows at the Royal Agricultural Society's meetings, were awarded to animals bred by Mr. Booth, of Killerby; in 1841, at Liverpool, to Bracelet; in 1842, at Bristol, to Necklace; in 1844, at Southampton, to Birthday; and in 1845, at Shrewsbury, to Ladythorn.

To this day it is a moot question, amongst those who remember the world-renowned twins, to which of them could be most justly awarded the palm of beauty. Necklace is said to have had neater fore-quarters, and to have been rather better filled up behind the shoulders. Bracelet had fuller, longer, and more level hind-quarters. Bracelet was the dam of the famous bull Buckingham, by Mussulman, and of Morning Star, by Raspberry, which was sold in 1844 to Louis Philippe. She also produced Birthday by Lord Stanley, whose career as a prize cow at the Royal, the Yorkshire, and the County of Durham Shows, was eminently successful. Bracelet was also the dam of Hamlet, by Leonard; and whilst in calf with him her stifle joint was dislocated, and, being incapacitated for further breeding, she was slaughtered. Birthday was born in the field opposite the house on the 20th of May, which happened to be Mrs. Booth's birthday. The herdsman seeing that Bracelet had calved, hastened to bring home the new arrival. The party sitting at dinner observed him, and turned out to welcome Lord Stanley's daughter, then approaching in a wheelbarrow, from which (and it was accepted as an augury of her future achievements, and an indication that such vehicle was all unworthy of its freight) she indignantly jumped, and staggering on to all fours, assumed a show-yard attitude. She was then and there unanimously ycleped "Birthday" in honour of the occasion, and on many happy returns of that day Birthday's health was drunk together with that of the amiable mistress of the mansion; and as her yearly recurring triumphs in the show-field were recounted, the episode of the wheelbarrow was not forgotten. Birthday, like her dam, was unfortunately injured in her stifle joint, and, after suffering for some time, and consequently wasting considerably, was slaughtered, when the flesh over her back and loin was found to be 5½ inches deep, and her total weight of beef 87 stones, of 14 lbs. to the stone. Birthday bred Gem, Genuine, Birthright, Lord George, and Brigadier. The two first were supremely beautiful. Gem was described by Mr. Booth as having been much like Queen of the Ocean, and possessing very sweet fore-quarters. It is said that Sir Charles Knightley—and England could boast no higher authority, nor one more studious of fair proportion—considered her the *ne plus ultra* of shorthorn excellence. In a volume of the "Herd Book" at Killerby is a slip of newspaper with the following paragraph: "At the Yorkshire Agricultural Show held this week at Wakefield, six pure bred Shorthorns were exhibited for competition by Mr. Booth, of Killerby, and each gained a first prize in the different classes in which they competed. We question much if such another exhibition could be produced by any other breeder of Shorthorns in the kingdom." This was in 1846. The following are the animals referred to: First prize yearling bull, Hamlet; first prize cow, Mantalini; second ditto, Alba; first prize three-year-old cow, Gem; first prize calf, Bloom; first prize in extra stock, Birthday. The newspaper version, it will be seen, is not quite accurate, for Bloom was exhibited by Mr. Richard Booth; and one gained only a second prize. She was, however, beaten by one of her companions.

Another daughter of Bracelet was Pearl, by Leonard, whose grandaughter Pearly was purchased at the Killerby sale by Col. Towneley. Pearly was by Royal Buck, who was by Buckingham, a son of Bracelet; her dam was Manille by Brigadier, who was by Morning Star, a son of Bracelet, and out of Birthday, a daughter of Bracelet; and her grandam was Pearl, a daughter of Bracelet. The close in-breeding of Bracelet herself has already been shown. Pearly was the parent of Ringlet, sold for 500 guineas to Mr. Douglas. She subsequently produced Pride, Frederick's Bracelet, Pearl, and Precious Stone.

Bracelet's twin-sister, Necklace, has unfortunately now no female representatives. She bred Diamond by Rubens, Stanley by Lord Stanley, sold to M. St. Marie, and Jewel, the dam of Jeweller. This bull was used in the Towneley herd, and was the sire of the famous cow Butterfly, and of many other prize animals. Jeweller was by Hamlet, who was Bracelet's son by Leonard; his dam was Jewel by Leonard; his grandam was Necklace, Bracelet's twin-sister; so that it is most apparent, even if this had been all, how much the Towneley herd was indebted to the Booth element which it contained for its great success. But this is not all; for—not to mention others—Valiant (12253), Master Butterfly the 4th (14920), and Butterfly's Nephew (15714), which were all used in the herd, were all of Booth families, the first of them being also principally of Booth blood, and the others containing a large amount of it, while a fair proportion of the young bulls and heifers owe their origin to Baron Hopewell, a pure Booth bull, of the Mantalini family, bred by Mr. Barnes. Jeweller was a small, rather shabby-looking bull, but with great grazing propensity. After the expiration of his first season at Towneley, he was returned to Killerby, and Mr. Booth fed him off.

Just after his sale and delivery to the butcher, Mr. Eastwood drove over to Killerby, and, walking round the yards, inquired for his shabby friend. "He has gone to the butcher," was the reply. "When? I have come to buy him," exclaimed Mr. Eastwood. "How could you think of sacrificing such a bull?" An express was sent off to arrest the axe, but unfortunately it was too late. It may be here observed that Mr. Eastwood at that time superintended the management of the Towneley herd, as well as of the estates, having first, by the sale to Col. Towneley of his own well-selected herd, and then by the excellent and judicious purchases which he made, laid well and skilfully that foundation on which Mr. Culshaw subsequently erected so grand a superstructure.

The Gaudy, alias Lady Betty tribe, originated in a cow bought of Mr. Marmaduke Taylor of Catterick, whose daughter's *sobriquet*, "Lady Betty," was bestowed upon the new purchase. Mr. Booth described her as a very fine cow, "short in her hair, but as round as a roller, without a hill or a hole in her." She was the dam of a cow very famous in her day, Old Gaudy by Suworrow; fifth in descent from whom came Madeline, the dam of Col. Towneley's bull Hudibras, and his superb cow Alice. The latter had perhaps few, if any, equals in her day. Her back was of astonishing width and levelness, her loin and chine being remarkably wide, and her ribs boldly sprung. Her fore-quarters and whole frame were very symmetrical, with the exception that her tail was set on rather high, and her "tuts" somewhat deformed with fat. She won the *Irish Farmer's Gazette* cup, value £120, in 1853, and the first prize and gold medal at Smithfield in 1854.

Towneley could formerly boast of the representatives of another family of the Killerby Shorthorns descended from the cow Floranthe, previously mentioned, as one of Mr. J. Booth's breeding stock in 1835. They were descended from Mr. Charge's cow Exhalation by Hulton. Floranthe's daughter Mantle, by Marcus, was sold by Mr. Booth to Mr. Bannerman, and by him to Col. Towneley. Mantle was the dam of Valiant, and the splendid prize cow Beauty, both by Victor, of the Bracelet tribe. Beauty was a fine, round-barrelled cow, of a rich roan colour. She had rather large, but well cushioned hips, exceedingly good breast, and very heavy top, but rather light thighs and flank. She won the first prize of £20 at the Royal Society's Show at Lincoln in 1854. She was the dam of Beauty's Butterfly and Beauty 3rd, whose son, Master Butterfly 4th, was sold to the Emperor of the French. Beauty's Butterfly was one of the most beautiful of shorthorn cows; her lovely head, with its open curly brows, and delicately chiselled muzzle, its waxy crescent horns and full lustrous eye; her gracefully swelling neck, well-sloped shoulders and broad full breast; her astonishing girth and rotundity of frame, and the even development of muscle over every part of it, composed a *coup d'œil* not easily forgotten. Her transcendent beauty, when she was shown in Baker-street, roused for a moment even Cockneys from the phlegmatic apathy with which they usually regard all pastoral subjects: she was the talk of " the town;" and *Punch* immortalized her in the elegant lines he put into " Joe Culshaw's" mouth. To a critical eye, however, Beauty's Butterfly was not exempt from failings: her hindquarters rather wanted length, and her thighs and twist fulness, to be in strict harmony with her otherwise perfect proportions. Entomologists speak of the "Twenty-plume Butterfly," a prefix which, I think, such an adept in nomenclature as Mr. Culshaw might appropriately have bestowed on this Houri of the herd, in token of some score of victories achieved by her. The term "Butterfly," it is true, has been objected to altogether by some, as conveying the idea of an airy, volatile, sportive ephemera, rather than of a long-faced, ponderous, bellowing beast of the field; but we have the "Long-horn Moth," why not the "Shorthorn Butterfly?" For my own part, I never penetrated into that Shorthorn Elysium, Towneley Park Farm, and witnessed the comfortable quarters and comfortable case of its inmates, without echoing, with slight modification, the aspiration of the old song—

"I'd be a Butterfly, born in a *Byre!*"

The Mantalini tribe, which has since attained a European fame, came to Killerby from Cleasby by the purchase from Miss Wright, Mrs. John Booth's sister, of Sylph by Remus, and it is a singular circumstance that when it came to Killerby the family had already passed through the hands of two of those four spirited breeders who jointly purchased at Mr. C. Colling's sale the bull Comet for the princely sum of 1000 guineas— Colonel Trotter and Mr. Wright. In Colonel Trotter's hands this family go back to those very early times of which all records are lost; and this probably accounts for the circumstance that while, in modern times, the pedigree of the Mantilinis concludes with "Strawberry by Son of Favourite (252), dam Strawberry," in all the early pedigrees of Colonel Trotter's cattle the dam of Strawberry by Son of Favourite is called "Hollon." In one of them she is said to be "Hollon, bred by Mr. Hollon, of Strepans, near Darlington;" and, in another, "Hollon, by a son of (Mr. Charge's) Dalton Duke (188)," which connects them with the third of the four purchasers of Comet. As might have been expected in cattle coming out of such hands, Mr. J. Booth's new purchase, Sylph, was full to overflowing of the blood of Favourite. Her grandam, Alpine, and her great grandam, Strawberry, were both by sons of Favourite; her dam, Matilda, was by a son of Comet, who was both son and grandson of Favourite, while she was herself by Remus, whose sire was not only a son of Comet, but whose dam and grandam were both by sons of Favourite. It can be no cause for surprise that such a family, descended from such blood, and crossed by such bulls as Pilot, Alderman, Matchem, and Marcus, gave origin to a cow so splendid as Mantalini, the winner of twelve first-class prizes at four years old. Mantalini, though on a less scale than Bracelet or Charity, was a very true-made cow, on whom it would have been difficult for any one to make a disparaging criticism. Though wanting, perhaps, somewhat of the imposing grandeur of the Isabellas, or the massive development of Queen of the Ocean, she was a beautiful animal, and an excellent dairy cow. Nor, though Warlaby and Killerby see them now no more, is the fame of this noble family yet extinct. Westland, the Killerby of the Sister Isle, can boast of many worthy representatives of it in the descendants of Modish and Milliner, a sister and daughter of Mantalini, purchased by Mr.

Barnes of the late Mr. J. Booth; and from Mantalini's daughter Pelerine (sold with her twin sister Polka to Colonel La Touche) came Mr. Douglas's three Graces, Rose of Autumn, Rose of Summer, and Rose of Athelstane, of whom it is sufficient to say that Rose of Summer had the unparalleled honour of winning the first prizes at the three national shows in 1854, and her daughter Rose of Athelstane in 1857. Seventh in descent from Sylph, and third from Mantalini, the beautiful cow Farewell by Royal Buck, then a yearling heifer, was sold at Mr. J. Booth's sale in 1852. She lived to the age of fourteen years, during most of which she was distinguished for her great milking powers; and she was the parent of ten live calves. The greater part of these were born at Stackhouse, where her daughter Claribel by Valasco, and her grandaughters, still carry on the line. Nor have her male descendants been less celebrated: of these Imperial Windsor, the sire of Mr. Foljambe's numerous prize-winners, and Mr. Ambler's well known prize bull Prince Talleyrand, the son of her daughter La Vallière, deserve especial notice.

From another cow of Mr. J. Booth's, called Belinda, sprang Young Matchem (2282), and from her grandaughter May Day, the bull Marcus. The latter, after doing good service at Killerby, as may be seen in the pedigrees of many of its best families, was sold with a cow called Yorkshire Jenny to the Rev. John Bolden to go to Australia. Unfortunately they were lost in a storm. Yorkshire Jenny appears to have had no pedigree except the cross of Priam. She was the winner of the second prize in the cow class at Northallerton in 1843.

Another cow, called Rubicon, by Priam, appears to have had numerous descendants, now dispersed. Mother Red Cap, which was sold to Mr. Fawkes, and her son Red Knight, winner of the first prize for two-year-old bulls at Lewes, were descended from her.

The Belinda and Mother Red Cap tribes seem to have originated, like the Mantalini and Floranthe families, in the stocks of Miss Wright and Mr. Charge.

It is curious to trace the family connection that existed between the first improvers of the Shorthorns. Mrs. Thomas Booth was a sister of Major Bower. Mr. Wright, one of the purchasers of Comet, was the father of Mrs. John Booth, and brother-in-law of Mr. Charge; and the last gentleman's sister married Mr. Colling of White House. In the Booth family alone the taste for shorthorns seems to have been an hereditary predilection. Whilst the names of nearly all the earlier breeders have disappeared from the Herd Book, that of Booth is still conspicuous; and part of the old Killerby and Warlaby herd is now in the hands of the third generation.

It has been observed that the bull Matchem appears as the sire of some of the cows in the Killerby herd. This is the same Matchem that Mr. Bates's Oxford tribe derive from. He was purchased at Mr. Mason's sale, by Mr. John Booth and Mr. Maynard of Harlsey. On looking over the pedigrees of the Booth cattle, however, it will be seen that a bull of alien blood has never been used *throughout* the herd; for Matchem and Exquisite were only very sparingly employed, but one or more of the best cows has been crossed with the bull selected, and the male produce made use of. If the progeny resulting from this half-cross proved unsatisfactory, they were either disposed of, or subjected to a course of extra close interbreeding. Of the early experiments of this nature the cross of Matchem appears to have been the most successful, and of the more recent ones the crosses of Lord Lieutenant and Mussulman. The former of these was adopted by Mr. Richard Booth, in the year 1839, for his cow White Strawberry, and the latter, in the same year, by Mr. John Booth, who sent his famous cow Bracelet to Colonel Cradock's Mussulman, of the grand " Old Cherry " blood, and a son of the famous old cow by Pirate, who bore that name. Good as that blood was, Mr. J. Booth does not appear to have introduced that cross with a view to engrafting on his herd any particular excellence exhibited by Mussulman which was wanting in his own cattle, but rather to have been acting in conformity with the general practice of cattle-breeders of that day, who seemed to think it a matter of course to effect, at certain intervals, some change or renewal of strain in their stock. There is a prevalent idea that the length of hind-quarter which distinguishes the Warlaby cattle was derived from Mussulman, but for this there seems to be no sufficient foundation. The lengthy hind-quarters had long been a characteristic of the Booth herds. The early bulls, Rockingham, Sir Henry, and Julius Cæsar, and the cows of that time, especially Gaudy, could not be excelled in that point by any animal of the present day. Bracelet, as many will remember, possessed this length of quarter in a remarkable degree; so also did her grandaughter Gem. It was sufficient for Mr. J. Booth that he saw in Mussulman, not merely the fortuitous possessor of equally valuable properties with those which his own herd could boast, but the inheritor of them; for being descended from the stocks of Messrs. Wright and Charge, who, as we have seen, were of the family coterie of shorthorn breeders, Mussulman's ancestry had all been well known to Mr. Booth for generations.

It has been asserted by *over*-zealous advocates of the system of close interbreeding, that the crosses of Mussulman, Lord Lieutenant, Matchem, and others, introduced scarcely any fresh blood into the Booth herds; for inasmuch as no alien bulls were used but those whose veins were surcharged with the blood of Favourite, the recourse to them was nothing more than a recurrence to, or renewal of, the old family strain; but this is really only what is true of every well-bred shorthorn of the period, and therefore proves nothing. Take any one of them, and trace back the pedigree of each of its progenitors (whose numbers of course increase each generation back in a geometrical progression), and this bull Favourite will be found to recur directly and indirectly a surprising number of times. The following elaborate calculations, for which I am indebted to the Rev. J. Storer of Hellidon, may be quoted in illustration of this: —Mussulman is 64 times descended from Favourite; namely, through Magnum Bonum 30, through Pirate 22, through Houghton 9, through Marshal Blucher 3; total 64 times. Lord-Lieutenant was 106 times descended from Favourite, and Matchem 52 times. Crown Prince is 1,055 times descended from Favourite, and Red Rose by Harbinger 1,344 times. So the produce of the two are descended from him 2,399 times. But work out the

Duchesses or any shorthorns of good blood, and the result will be found very much the same. It will not do, therefore, to claim bulls as of kindred blood on this ground only. Moreover, it must in candour be admitted by the advocates of in-and-in-breeding that a careful consideration of the above facts leads to one unavoidable conclusion. Very strong in-and-in-breeding is a totally different thing in our case from what it was in the case of the earlier breeders, the Collings and Mr. Thomas Booth—so different that there can be but little analogy between the two cases. They bred in-and-in from animals which had little or no previous affinity. We breed in-and-in from animals full of the same blood to begin with. In our case the *via media*, and therefore the *via salutis*, would seem to lie in the adoption of two apparently opposite principles—*in-and-in-breeding*, and *fresh blood*. It is manifest, however, that this latter principle should be acted upon with extreme caution, or to a very limited extent, when it is desirable to preserve and perpetuate the distinctive type of any particular tribe, especially when, as in the Warlaby herd, there is no visible deterioration in symmetry, substance, or stamina, or any want of fertility traceable to in-and-in-breeding. Yet even in such cases it is doubtless advisable to have occasional recourse to remote alliances, taking care to have as many removes as possible between members of the same family ; or, where using bulls nearly related to the cows, giving the preference to such as have been subjected to different conditions of life, it being a well-known physiological fact that a change of soil and climate effects perhaps almost as great a change in the constitution as would result from an infusion of other blood.

In July, 1852, the Killerby herd was sold by auction; and the sale was attended by breeders from all parts of the kingdom. There was unfortunately, however, at this time, a general depression in the value of all agricultural produce, and the cattle did not realize prices at all adequate to their merit and celebrity. Some of them have since been resold for three times as much as they fetched at the sale. Venus Victrix, the daughter of Bloom, brought the highest price. She was purchased by the late Mr. R. Booth for 175 guineas, and presented to his brother. In the report of the sale in the *Mark Lane Express* it is stated that, "since the formation of the Royal Agricultural Society of England, in 1839, the prize for the best cows has been gained either by the Messrs. Booth or by animals bred from their stock, with the exception of the two first meetings, when they did not exhibit, and that of 1843. The animals which took the prizes on the two former occasions were both afterwards exhibited against, and defeated by, Bracelet."

After the dispersion of the herd, Mr. J. Booth entered only very partially into the breeding of shorthorns. Venus Victrix's first calf died. She afterwards produced the two well-known bulls King Arthur and King Alfred, both by Crown Prince—the sires of many a noble Shorthorn, and the earners of many a golden guinea for the treasury at Killerby—and two as beautiful heifers, Victrix by Royal Buck, and Venus de Medicis by Harbinger; after which she fell into one of the ditches at Warlaby, whither she had been sent to Crown Prince, and becoming partially paralysed in her hind-quarters, was fed off.

Venus Victrix was a cow of exceedingly good form. Her back was broad and level, her ribs well arched, her breast heavy and wide, and her thighs full. Her weak point was that which is usually found in conjunction with good milking qualities—a little want of substance in the neck and chine. She won five prizes at the Royal and Yorkshire shows. Her daughter Victrix gave birth to Venus Astarte, a lovely heifer by Baron Warlaby, which did not breed, but took the first prize for extra stock at the fat show in York in 1861. Victrix's next calf was retained for fifteen months *in utero;* and after getting rid of it she fell in-calf again to Windsor, and aborted, when all hope of her breeding being at an end she was slaughtered.

Venus de Medicis was a very symmetrical heifer. She was of great breadth, was thickly fleshed, and had very finely-finished quarters. She was sold to Mr. Douglas for 300 guineas, and sacrificed to the Goddess of Display. After winning several prizes, she cast a calf at the Paris Exhibition; and so ended her chance of further increase.

In proof of the fluctuations to which, from various causes, the procreative power of high bred bulls is subject, it may be here mentioned that Venus's son, King Arthur, soon after being hired for the first time, was sent home as unfruitful. Shortly afterwards he was re-let to Mr. Bruere, and became, in the first year, the sire of eighteen calves. He was again let into Ireland, and there begot only two calves. He returned home, was re-let to Messrs. Wood and Lawson, and gave entire satisfaction. Mr. R. Booth's Prince George, half-brother of Monk, affords another instance of this. He was returned to Warlaby as useless. Mr. Bruere had been so successful with King Arthur under similar circumstances, that Prince George was sent to him, King Arthur's term having just expired. In his first season at Mr. Bruere's Prince George was the sire of twenty-eight calves.

It was matter of frequent experience with Mr. Booth that bulls were sent back to Warlaby as unfruitful, which yet impregnated cows the very week of their return, and afterwards proved very useful in some other herd. It is a fair inference from this fact, that the fault in these cases lies quite as much in the herdsman, the cows, the situation, or the diet, as in the bull. To study the peculiar temperament of these sultans of the herd is not the least important task

"That to the faithful herdsman's art belongs."

In one bull pride predominates—his dignity must be respected—he must not be hurried or struck: in another modesty prevails—he must not be watched; one is apathetic, and requires to be roused to exertion; another is timid, and encouragement is necessary; one requires a generous regimen and comfortable housing; another (and this is true of most bulls) requires natural food, grass, hay, and turnips, and, above all things, that which is indispensable to the vigour and potency of all animal life —daily exercise.

Though the great Killerby Herd had been dispersed,

and its queenly matrons consigned to other hands, the family taste for shorthorns lingered still, like a presiding genius, on the accustomed spot. A few good animals of shorter but distinguished lineage were retained, and these have given rise to other families, which, if more recent in their origin, are proud to trace that origin to the hands of the Booths, and the quiet meadows of old Killerby. Some of these were disposed of by the late Mr. John Booth, while he yet lived, and several of them came to Stackhouse.

One of these importations from the Valley of the Swale to that of the Ribble, was Calomel, purchased when an aged cow, with her bull calf, for 150 guineas. She was a very useful white cow by Hamlet, tracing back, through dams by Leonard and by Buckingham, to the once celebrated herd of Sir Matthew White Ridley. She bred several valuable things at Stackhouse, and her daughters and granddaughters yet remain there, and in the herds of Mr. Aylmer of West Dereham, and Mr. Storer of Hellidon, whose fine cow Modred is the daughter of another excellent shorthorn which Killerby supplied to Stackhouse, Mistress Mary by Baron Warlaby, the other bulls in her pedigree being Royal Buck, Hopewell, and Hamlet.

Of the cattle which remained at Killerby, and descended at Mr. J. Booth's death in 1857 to his sons, I have previously mentioned Venus Victrix and her family: they were the *crême de la crême* of the herd. Another most valuable cow, though of more recent origin, remained at Killerby at Mr. Booth's death—Hecuba, by Hopewell, from a dam by Hamlet, out of a Leonard cow. This cow has all the Booth character, and transmits it to her offspring. She is in colour a dense red; a large animal on short legs; when not in milk, laying on flesh with wonderful rapidity; and when in milk, she is what every cow ought to be, a great and deep milker, with an udder whose size and form might provoke the cupidity of even a London dairyman. When young she was pronounced by one of the best of judges, Mr. Eastwood, to be the very type of the true shorthorn, the very model of a bull breeder. Hecuba, though born in the year 1851, and therefore now about 16 years old, is as fresh as an eight-year-old cow, and had a living calf last September. Besides several bull calves, one of which is Trojan Warrior by Windsor, she has bred eight heifers, and many of them promise to be equally productive. It is impossible to enumerate all of them, still more so to mention their excellences, but her daughters Forest Queen by Royal Buck, and Queen of Trumps by Welcome Guest are well known to all who have visited Killerby, while her younger daughters and many of her grandaughters by Valasco are of at least equal promise.

But the pearl of pearls, the most valuable cow in the present Killerby herd, remains to be mentioned last, Soldier's Dream by Windsor, from Soldier's Nurse by Crown Prince. Descended from the Moss Rose family, a daughter of the unrivalled Windsor, and of a mother who was own sister of Prince Alfred of Imperial and Royal fame, Soldier's Dream is everything that might be expected from her lineage, and gives fair promise of shortly transmitting her perfections. Her dam Soldier's Nurse was a present from the late Mr. R. Booth to his nephews when they lived together at Killerby, and was second to few even in Warlaby's distinguished herd. She was considered by many good judges, before she left Warlaby, to be superior to even Queen of the Vale; but Mr. Booth's practised eye saw the difference, and she must be placed after, though at no long interval, the unrivalled Queen. She was not quite so large, and rather darker in the red. She inherited from her mother an almost Bride Elect breast, was beautifully ribbed up, and, with faultless hair and touch, she was the very picture of a shorthorn.

Besides Soldier's Dream she left behind her the very valuable, thick, heavy fleshed bull Brigade Major by Valasco.

THE WARLABY HERD.

It is now necessary to take a retrospect of the herd at Warlaby, commencing with the year 1835, when Mr. Richard Booth, inheriting the estate, went to reside there. Mr. Booth's residence at Warlaby is a modest, unassuming, country house. It stands environed by well-timbered paddocks in a rich meadowy tract of country bounded by distant hills, and known as the Vale of the Wiske. It is one mile from the village of Ainderby, of which it is a hamlet, and about three from Northallerton, the central town of the North-Riding. The farm as occupied by Mr. Booth consisted of 310 acres, about half in pasture; other farms then let off, have since his death been added to it. The land is better in character than that at Killerby; it is chiefly clayey loam, and grows fine wheat and turnips, and long hay. The pastures are well adapted for cows, but unsuited for sheep, because liable to be flooded. The River Wiske, which still retains its Gaelic name, Uisg (water), being the most sluggish of all the North Yorkshire brooks, and having the shallowest stream-channel, frequently overflows the lower pastures, and large deep ditches, which have been fatal to many a good cow, intersect the fields to carry off the water.

The house was everything that an old bachelor or his friends could require; and many a visitor there can bear testimony that within its walls reigned supreme the open-hearted northern hospitality to an extent that Southrons know not. Many a valuable cup and hard-won medal may there be seen; the portrait of many a prize-taker decorates its rooms; and many a pleasant hour has been spent and ancient story told in that quiet Shorthorn home, while the genuine old Squire

> Refilled his pipe, "and showed how fields were won."

Shortly after settling at Warlaby, Mr. Richard Booth had quite made up his mind to give up the breeding of shorthorns, and had already sold individual animals from the Strawberry and Moss Rose tribes, when a bantering remark made by a gentleman in the neighbourhood, to the effect that "the Booths had lost their Blood," incited him to change his purpose, and put his friend's assertion to the proof. The Warlaby herd had for some years past been kept very much in the shade, Mr. Thomas Booth having been latterly intent only on breeding useful animals, without aspiring to the honours, or courting the notoriety of public exhibition; but Mr. Richard Booth felt assured that it contained ample materials to enable him to guard the laurels that had been bequeathed to him.

THE BOOTH HERDS.

Amongst the cows then in the herd, and whose descendants have earned a wide-spread renown, were Strawberry 2nd by Young Alexander, and Rally by Rowton, both of the Halnaby or Strawberry tribe; Flora by Isaac, of the Farewell tribe; Blossom by Young Red Rover, and Young Red Rover himself, of the Blossom tribe; Broughton by Jerry, and her daughter Young Broughton by Young Matchem, of the Broughton tribe; Moss Rose by Priam, of the Dairymaid tribe; and Christon, also by Priam; to the number of which Mr. R. Booth now added one of the best cows of her day, his own Isabella by Pilot, then sixteen years old.

This cow Isabella, which has already been spoken of in the account of the Studley herd, produced at Warlaby Isabella Matchem, by Young Matchem, who gave birth to Fitz-Leonard by Leonard, Vanguard, Innocence, and Isabella Buckingham, all by Buckingham, and Isabella Exquisite by Exquisite. Fitz-Leonard was a neat, moderate-sized bull, standing on very short legs. What higher merit can be ascribed to him than that he was the sire of Crown Prince? Vanguard was a noble bull, as all the Buckingham bulls were. He was seven years on hire at Mr. Torr's, who, in 1853, exhibited him at the North Lincolnshire Agricultural Show, where he obtained the first prize of £20 (open to all England), beating several famous prize animals. Innocence was a small, level cow, with a luxuriant coat of hair. Her calf Leonidas by Leonard became a very fine bull, and was the sire of the notable Monk, one of the best of the Warlaby bulls. Leonidas's hair was so long that it waved about in the wind like the wool on a sheep's back. Isa-

E

bella Buckingham was a superb cow of great substance. She weighed when slaughtered upwards of 100 stones of 14 lbs., though she had been previously much reduced by the foot-and-mouth disease. She won seven prizes at the Royal, the Yorkshire, and the Highland Shows, including two firsts and a second (to Charity) at the Royal meetings. Isabella Exquisite by Exquisite was a cow of only ordinary appearance.

Isabella Buckingham was put to Exquisite; but the result of this cross, in the cow Sample, was equally unsatisfactory. She was a plain animal, with scarcely any of the Warlaby characteristics; nevertheless, her daughter, Isabella Hopewell by Hopewell, which was sold at the Killerby sale, under the name of Ecstasy, to Mr. Douglas, was a good cow, and a very good breeder. Another daughter of Sample's—Specimen by Crown Prince—was herself put to Crown Prince three times, and produced the prize cow Lady Grace and the bulls Sir Colin and The Corsair, in all of whom the Warlaby type was completely restored. The Corsair was sold to Mr. McDougall of Australia, but unfortunately died in crossing the Line. Lady Grace took the first prize for the best cow in calf or milk at the Cleveland Agricultural Show at Yarm in 1861, and was thus referred to in the report of the Show in the *Mark Lane Express*: "Mr. Booth's Lady Grace is a remarkable specimen, and a very fair specimen, of what a shorthorn of Mr. Booth's blood really is, in its natural state. Although she has gone to grass with ordinary kine, and been fed as any ordinary cattle are and should be, she is nevertheless exceedingly fleshy, and displays all those various points of excellence, particularly about the shoulder and quarter bones, covered as they are with a rich coating of elastic meat, for which the Booth breed are distinguished." Lady Grace, a daughter and grandaughter, as we have seen, of Crown Prince, was herself put to a son of Crown Prince—Prince Alfred—and the offspring, Graceful, was all that her name implies. She was one of the pair of heifers that took the first prize at the Worcester Royal, where she unfortunately cast twin calves. Lady Grace and her daughter Graceful, notwithstanding their in-and-in breeding, showed every indication of the most robust constitution, and had abundance of hair.

Whilst entirely concurring with Mr. Booth in the opinion that, as a rule, family alliances have been followed by more favourable results in the Warlaby herd than the resort to extraneous blood, yet I should suggest that it has been too much the fashion amongst "Booth men" to form conclusions unfavourable to such bulls as Exquisite, Water King, and Lord Stanley, from the acknowledged shortcomings of their *immediate* offspring. It is usual, for instance, to cry down Exquisite, and to say that the use of him was unsatisfactory. It undoubtedly was so, *in the first instance*, in more cases than those above mentioned; and why? Because the Booths deviated from their usual course, and at one stroke introduced *too much fresh blood*. They had previously introduced the fresh blood of Mussulman, but in a modified form, through Buckingham, and that of Lord-Lieutenant in an equally modified form through Leonard; but they infused the blood of Exquisite into the herd raw and undiluted. The result might have been foreseen.

But the question really is, not what was the *immediate* consequence of this overpowering dose of fresh blood, but, what were the results when the Exquisite blood became gradually more incorporated with the Booth in the same manner and with the same modification as that of Mussulman and Lord-Lieutenant?

The description of Lady Grace, just quoted from the *Mark Lane Express,* is a sufficient answer. Nor are there wanting other examples than those of Lady Grace and her blooming daughter to prove that the blood of Exquisite has done no more harm to the Booth cattle than did the blood of Mussulman when they have it in their veins in an equally diluted form. Not to mention others— British Hope, purchased by the late Mr. Langston at Lady Pigot's sale, was, in the opinion of good judges, the best Booth bull sold there; he was by British Prince from Isabella Hopewell, a grandaughter of Exquisite, and a cow which had previously bred, by a bull partially Booth and of a Booth family, the celebrated Lamp of Lothian.

Nothing could be more respectable than Exquisite's lineage, combining as he did the best of the Chilton and Wiseton blood; and I am certainly justified in imagining that, as an *animal,* he must have been first-rate, or three such unquestionable judges as the late Mr. John Booth, Mr. Richard Booth, and Mr. Torr would hardly have given 370 guineas for him at the Wiseton sale at thirteen months old, the highest price *then* on record of any animal of that age sold by public auction. Exquisite had plenty of substance, and a profusion of beautiful hair, a characteristic which may often be observed in his descendants.

In futher illustration of my argument, I may here refer to the fruit of the union between Mr. John Booth's world-renowned cow Bracelet and Col. Cradock's Mussulman—the bull Buckingham, whose name and fame will endure as long as shorthorns exist. To superficial observers, Buckingham was, as is well known, a plain, ungainly animal, the result of the first cross with Mussulman being apparently just the same as the first cross with Exquisite. In the hands of a less discerning judge Buckingham would probably have been steered; but his owner's practised eye, accustomed to weigh the relative values of points and proportions, and to detect latent capabilities in animals bovine and equine, saw beneath Buckingham's plain exterior his high qualifications for a sire.

Mr. Richard Booth showed an equal appreciation of the merits of this bull, and determined to secure him for his own herd. An entry in the margin of the late Mr. J. Booth's herd-book records the fact that he "sold Buckingham to his brother Richard for £150." It remained for the future to demonstrate the wisdom of this purchase. No visitor to Warlaby could ever appreciate the merits of this bull until his offspring proclaimed it. Never were calves with backs so broad, ribs so round, shoulders so shapely, flanks and fore-quarters so full and deep. Buckingham's career was unfortunately brief. Two distinguished breeders from the Emerald Isle, Mr. Barnes of Westland and a brother breeder, arrived at Warlaby in quest of a bull. After viewing the herd with

Mr. Booth, they retired to prepare for dinner, when Mr. Barnes remarked to his friend, "We must hire Buckingham." "You cannot be serious," replied his friend, "in proposing to hire such an ugly brute!" "Look at his stock," said Mr. Barnes. "If you hire him, you must hire him alone," rejoined his companion, "for I tell you candidly I cannot join you." "I will," said Mr. Barnes; and he did so. The tragical issue of this hiring—how the steamer conveying the bull caught fire in mid-channel, and with what characteristic devotion his Irish conductor shared the poor animal's funeral pyre on the burning deck of that ill-fated bark rather than desert his charge, are matters of history already chronicled. The loss of this bull was a national one, as an enumeration of some of the animals he bequeathed to Mr. Booth will show: Charity, Plum Blossom, Bloom, Bagatelle, Bonnet, Medora, Vivandière, Isabella Buckingham, Vanguard, Hopewell, Benedict, Baron Warlaby—all not only famous in themselves, but the parents of animals whose names are as familiar in our ears as household words.

Moss Rose by Priam, one of the cows which came into Mr. R. Booth's possession in 1835, was the grandam of Minette by Leonard, and of Mr. Mason Hopper's prize bull Master Belleville, the sire of Mr. Storer's Rosey, and through her grandsire of the Duke of Montrose's Rosedale. Minette was the dam of Vivandière and Royal Buck, both by Buckingham. Amidst the galaxy of prize beauties hereafter to be mentioned, the modest Vivandière, with her beautiful head, was frequently unobserved, except by the admirers of a well-filled udder, unless brought into notice by a quiet observation from her owner of, "Look at that head and hair!" She, however, amply vindicated Mr. Booth's judgment, and her own claim to something more than a passing notice. She had ten calves. One of her sons, Prince Alfred, was a very stylish bull, with just the mould and hair to ensure his success in the show-field. In his youth, however, he was only exhibited once—at Berwick-on-Tweed, where he won the first prize as the best bull-calf.

> "For other fortune then he did inquire,
>   And, leaving home, to Royal halls he sought,
> Where he did let himself for yearly hire,
>   And in the Prince's service daily wrought."

In other words, Prince Alfred had the honour of serving two years in the late Prince Consort's herd at Windsor, and one year in the model farm of the Emperor of the French at Fouilleuse; after which he had the further distinction of spending two years on the farm of that Queen of English Bucolics, Lady Pigot, who, like her sprightly and bewitchingly-entertaining exemplar Lady Wortley Montague, seeks repose from the occupations of the gay world in pastoral and agricultural pursuits. In 1864-5 Prince Alfred, though then a veteran in the decline of life, again entered the lists, and at least half-a-dozen times vanquished all opponents.

Vivandière produced, besides Prince Alfred, Welcome by Water King, Vivacity by Fitz-Leonard, Verity by Vanguard, and Campfollower and Soldier's Nurse, both by Crown Prince; also Prince Arthur and Prince Oscar by Crown Prince, and Knight Errant by Sir Samuel.

Water King, the sire of Welcome, was by Baron

Warlaby, his dam by Fourth Duke of Northumberland, g. d. Waterloo 3rd by Norfolk, g. g. d. by Waterloo, g. g. g. d. by Waterloo. He was put to a few other animals in the herd, but the cross not being thought to have amalgamated well with the Booth blood, except in the case of Peach Blossom and Water Nymph, his use was discontinued. Welcome, however, it must be admitted, furnishes another illustration of my argument, that, though the fruits of Mr. Booth's experiments in crossing were not at first such as to encourage him to depart from his usual course of adhering to proved blood, it is hardly fair to jump to the conclusion that these crosses were injurious, or even unproductive of good, because their immediate results were unsatisfactory. Welcome, it is true, was a homely enough cow herself; in fact, one of the old-fashioned, unimproved type; but she gave birth to a heifer, Welcome Hope by Hopewell, little inferior in style and merit to any animal in the herd. "Ye ken," quoth old Cuddy, "some folks say we sud gang for a change o' bull. Ye see we did gang in ould Welcome. Whya, she be a vara useful cow, and mony a yan has nae sae gude a yan; but is she to be mentioned i' t' same day as Campfollower? Nout o' t' kind. She munnot be i' t' same year. See ye at Campfollower, doesn't she walk away frae ye like a Queen!" "But her daughter, Welcome Hope," we remark, "is a good animal." "Aye! Hopewell has putten in some gude wark when he gat that there heifer. She wad mak' up a slashin' cow, though she *have* a lile touch o' Bates bluid in her; but then, ye ken, ould Hopewell wad mak' 'up a' deficiencies."

Vivacity by Fitz-Leonard, another daughter of Vivandière's, was sold by Mr. R. Booth to Mr. Bolden, together with Young Rachel, the dam of Mr. Ambler's Grand Turk, Bridget by Baron Warlaby, and another. They were purchased by Mr. S. E. Bolden, for his brother in Australia, just at the time the gold discovery caused a rush of emigrants to that country, and it was found impossible to ship them at anything like reasonable rates. Mr. Bolden informed Mr. Booth of this, and offered to return the heifers; but Mr. Booth gave him the option of retaining them himself, which he was glad to do. Vivacity was a strikingly beautiful heifer, but unfortunately died in giving birth to her first calf, May Duke by Grand Duke; which was purchased for the Stackhouse herd and afterwards sold to the Hon. Noel Hill, in whose herd, and that of Mr. Robinson of Clifton Pastures, whose property he subsequently became, he did valuable service.

Verity by Vanguard was the grandam of Sincerity, who stood second to Queen of the Ocean at Guisboro', and was one of the six animals that won the silver cup at Skipton. Her dam Truth died three weeks after calving, and Sincerity was brought up amongst the calves of the nurse cows, skimmed milk and porridge being "the best of her diet." She was calved in January 1859, and through the winters of 1860 and 1861 she never had a roof over her head. During the first winter Bridal Wreath, Dora, and Blush were her companions; and through the next, Rosette and Princess Elizabeth shared with these four the rigours of an unusually severe season. Sincerity was taken up just before calving in 1862, and after producing twins it was determined to "put her in training," as the slang phrase goes, for the show-yard. After her successes in the North, it was decided to enter her for the Worcester Royal Show, but she unfortunately displaced the patella of the stifle-joint, and was slaughtered. Sincerity had much of the character of Nectarine Blossom; perhaps was even more cylindrical in her proportions. Her fore-ribs were as round as a barrel.

Vivandière's daughter, Campfollower by Crown Prince, was a truly noble cow, with a queenly form and gait. She had capital crops, a broad flat back, and was very wide between the legs. When she and Lady Grace were heifers together, it was difficult to determine the point of superiority between them. She was the dam of Soldier's Bride by Windsor, General Hopewell by Hopewell, Soldier's Joy and Soldier's Daughter, both by Lord of the Valley, and of the splendid bull Commander-in-Chief by Valasco.

Soldier's Bride, with her broad, deeply-covered back and circular frame, her wonderfully expansive shoulders, girth, and bosom, is too well-known to require any description here; it is enough to say, that a slight droop of the hind-quarters forms

"The sole alloy of her most lovely mould."

At the Royal Agricultural Meeting at Worcester in 1863, she shared with Queen of the Ocean the honour of the first prize for the best pair of cows, and has won nine other prizes at the principal shows. When she was exhibited as a yearling at the Northern Counties Show at Darlington, where she won the Founder's cup as the best animal in the yard, her estimated dead-weight, she being then only one year and eleven months old, was 70 stones of 14 lbs.—an instance of early maturity of which there is no parallel on record. Her half-brother General Hopewell was one of the neatest and best-framed bulls I have ever seen at Warlaby.

Soldier's Nurse, Vivandière's fourth daughter, has already been spoken of in the account of the Killerby herd, in which she bore the bell.

Flora by Isaac, of the tribe then known as "the Hope tribe," but since distinguished as "the Farewell," was another of the cows which Mr. Booth inherited with his paternal fields. She was the dam of Farewell by Young Matchem, from whom came in successive generations the ever memorable cows Faith by Raspberry, Hope by Leonard, and Charity by Buckingham, besides being the ancestress, through Hope, of Harbinger and Hopewell, and through Charity of the most famous bull of recent times—Crown Prince. Faith was a large, fine, but rather masculine cow. She stood second to Necklace at Doncaster in 1843, and second at Richmond in 1844. Her daughter Hope by Leonard was a magnificent cow, some even venturing to affirm, better than Charity. She won five first prizes.

Of Charity, who so long graced the Warlaby pastures, it is sufficient to say that she was the personification of all that is beautiful in shorthorn shape. Such was her regularity of form, that a straight wand laid along her side longitudinally from the lower flank to the fore-arm and from the hips to the upper part of the shoulder blades, touched at almost every point; her quarters were so broad, her crops and shoulders so full, her ribs so boldly projected, and the space between them and the well-cushioned hips so arched over with flesh as to form a continuous line. It was difficult for the most hypercritical eye to detect a failing point in this perfectly-moulded animal, and it was in consequence of Mr. Booth's high appreciation of her merits and those of her son that he made such free use of Crown Prince. Charity won every prize for which she was shown save one, when she was beaten as a calf by another of the same herd, after which her career was one of unvaried success. She was thrice decked with the white rosette at the Royal and thrice at the Yorkshire meetings.

Hope's son, Harbinger, by Baron Warlaby, with the exception of a little too much prominency of hip, was a very first-rate bull. He was let for a higher rent than any previous bull had been—250 guineas a-year. He was the sire of Mr. Booth's prize cow Bridesmaid, and of Red Rose, the dam of the incomparable Queens. Harbinger won the first prize in the bull-calf class at the Royal at Exeter in 1850, the only time he was shown.

Hopewell, another son of Hope's, and by Buckingham, was a fine lengthy bull, with a grand head and crest, and remarkably soft well-covered hips. He was exhibited at the Royal at Norwich in 1849, and with the whole class in which he stood, was condemned as not worthy of a place. At the Yorkshire show at Leeds the same year he was first in his class, and subsequently made nearly £1000 in hire.

Charity's world-renowned son, Crown Prince, needs no panegyric here. To the visitor at Warlaby I would say: "Si monumentum quæris, circumspice." Nor are the memorials of his achievements confined to Warlaby: they may be traced in a hundred herds to which his sons have transmitted the virtues of their sire. Crown Prince was never exhibited; but Mr. Booth truly said of him that "though he was not a prize-winner he was a prize-*getter*," an assertion which the records of almost every important showfield in the kingdom will verify. At one time Mr. Booth had sixteen bulls by him, let at once, at from 100 to 250 guineas each (his total number out at hire being then 26). He refused an offer of 300 guineas for a year's use of him in the Stackhouse herd, and a similar offer, it is said, from Aylesby.

Fame—a sister of Faith—which was sold to Mr. Walker of Maunby, produced Fanchette by Petrarch, Fay by Foigh-a-Ballagh, and Florence by 2nd Duke of York. From Fanchette are descended Mr. Torr's Fairmaid by Usurer, Mr. Pawlett's Fairy by The Corsair, and Mr. Barclay's Faith by Sir Charles. Four of this family, two cows and two heifers, were sold at Mr. Sanday's sale in 1861 for the large sum of 675 guineas. Fay produced Mr. Bolden's Fenellas by Grand Duke; and Florence was the dam of that gentleman's Duke of Bolton by Grand Duke, and of Little Red Rose by Petrarch. The latter, which was sold by Mr. Bolden to Mr. Douglas, was the dam of Norma, who has, in the hands of Mr. Wood of Castle Grove, given rise to a numerous and most valuable family, one of which is Mr. Wood's splendid cow Coquette, whose daughters, Clarionette and Castanet, have not been without renown in English showfields. *Bell's Weekly Messenger*, in a notice of the Castle Grove herd, mentions the descendants from Norma as " Mr. Wood's favourite family, 'the best he ever had;'" and gives the following particulars in relation to them:—" Fame, Farewell's daughter by Raspberry, having been put to the Bates bull 2nd Duke of York (5959), 'a beautiful little bull,' the son of Duke of Northumberland and Duchess 41 by Belvedere, produced, in the hands of Mr. Carruthers, the large and noble cow Florence, in whom the Booth blood of her predecessors received its first dilution, to the extent of one half. Whilst in the hands of Mr. Douglas of Athelstaneford, Florence calved Little Red Rose by Mr. Whitaker's Petrarch, 'an extraordinary good bull,' according to competent judges who knew him, and a quarter Booth; and from Little Red Rose Sir H. H. Bruce bred Norma by Druid (10140), a son of Baron Warlaby. In the person of Norma, the family came into the possession of Mr. J. Grove Wood, and her first female offspring at Castle Grove was Coquette, got by Comet (11298), a son of the Usurer bull Broker (9993), and out of Fair Frances by Sir Thomas Fairfax. Beyond Coquette by Comet the additions are due to Mr. Wood, and they indicate his leading partialities. The character of the original family blood is restored; the course so long deviated from is resumed; Prince Arthur (13497) was used, and King Arthur (13110); then Sir Roger (16991), Elfin King (17796), Sir James (16980), and, lastly, British Crown (21322), a son of Lord of the Valley and Bridal Wreath by Crown Prince. Some of these sires were used to some of the tribe, and some to others. Three crosses are the greatest number added by Mr. Wood to the pedigrees of Norma and Coquette. It is but a small number *in sound*, or to *the eye;* but it makes, when occurring as it does in these and similar cases, no less than seven-eighths of an animal's ancestry."

Farewell, after giving birth to the above mentioned produce at Warlaby, was sold by Mr. Booth to Mr. Walker of Maunby. It seems singular that Mr. Booth should have parted with so very valuable a cow; but probably he did not then anticipate the world-wide fame which awaited her descendants. In Mr. Walker's hands she gave birth to several calves; but two only of these have maintained the family. These were her twin-daughters Clementina and Clematis by Clementi (3399). The well-known prize bull Clementi was bred by Mr. Parkinson, and, like his own and equally celebrated brother Collard, was half Booth, being a son of Mr. Booth's Cossack (1880), from Mr. Shafto's famous cow Cassandra by Miracle. Of his twin-daughters from Farewell, Clementina and her descendants were unfortunately crossed with other blood, but have produced valuable animals in the herds of Mr. Foljambe of Osberton and Mr. Lynn of Stroxton.

Fortunately, Farewell's other twin-daughter, Clematis, met with a better fate; and her descendants by Booth bulls continue the royal line. I say fortunately; for, alas! the Farewell family are extinct in the female line at Warlaby; and if it were not for Clematis and her descendants, we should look in vain for pure Booth animals which trace their origin to Farewell. While yet at Mr. Walker's, Clematis gave birth to Baroness by Baron Warlaby; and Baroness, after passing through the hands of Mr. Harvey Combe, became, while the property of Mr. Crawley, the dam of Barmaid by British Prince.

Barmaid, a large framed massive cow with much Booth character, but also with something of the masculine character which was observed in Faith, was transferred to the Stackhouse herd, and there was fortunately put to Valasco so celebrated for his excellent heifers. The bull exactly suited her, and the produce was Dame Quickly—now happily no longer the last representative of the old Farewells; for two fair scions have sprung from her, Dame Patience and Dame Margery, and these, like Faith and Hope revived, give fair promise of continuing a family so celebrated; while their dam's value is attested by the post her son, Knight of the Garter, fills in the eminent herd of Mr. Foljambe. Dame Quickly is pure Booth, with the exception of a remote half-cross, which causes one thirty-second part of her blood to be that of the famous cow Cassandra—so that this family may fairly claim the first place as representatives of that noble line which no longer perpetuates itself at Warlaby.

The Strawberry or Halnaby tribe next claims our attention. We have seen that on Mr. Richard Booth's accession to the herd in 1835, it contained two cows of this family, Strawberry 2nd by Young Alexander, and Rally by Rowton.

Third in descent from Rally came Young Rachel by Leonard, previously mentioned as one of the four sold to Mr. Bolden. She gave birth while at Springfield to the lordly Grand Turk by Grand Duke, afterwards the pride and glory of Mr. Dodds and the Watkinson Hall herd.

Strawberry 2nd by Young Alexander, dam Strawberry by Pilot, grandam Strawberry by the Lame Bull, was dam of the famous cow White Strawberry by Rockingham. This cow, White Strawberry, was bred in every direction from the closest affinities of blood, her ancestors male and female being filled by repeated crosses with the blood of Albion and Pilot. She was a magnificent, broad-backed, wide-breasted animal, quite equal in merit to Anna and Isabella by Pilot, the two best cows that any of the three herds had previously to 1835 produced.

White Strawberry was the dam of Leonard by Lord Lieutenant, and the ancestress of that branch of the Halnabys that produced Monica by Raspberry, and her offspring Prince George by Crown Prince, Monk by Leonidas, and Modesty, Medora, and British Queen, by Buckingham. Monk, a magnificent bull, was absent from Warlaby on hire for ten years, and Prince George for nearly as long. Modesty, though the twin sister of the handsome Medora, was not much in herself to boast of, unless the old adage be applied to her of "handsome is that handsome does." She had twelve calves, among which were the well-known animals Water Nymph, Chastity, and Majestic. Water Nymph did not breed, and was fed off. Though entirely grass-fed, never having been in the house summer or winter, she won the first prize as the best fat heifer at the Yorkshire show in 1856. Chastity, but for the one failing of a rather upright shoulder, was a very superior cow of true proportions. She was the dam of Sir Roger by Windsor. She was drowned in the Wiske owing to the ice on which she had ventured giving way. Modesty's two last calves were heifers by Lord of the Valley: the elder one, Maiden's Blush, is a full-haired, short-legged, deep-breasted cow, though she has had, as Cuddy says, "nobbut a poorish putting on."

Modesty's twin sister Medora was a very stylish heifer, of great promise, which she did not live to fulfil; for

> "She, to give the *herd* increase,
> Forfeited her own life's lease,"

and died after giving birth to her first calf, Red Rose by Harbinger.

No cow ever trod the Warlaby pastures with a more becoming grace than Red Rose. With the exception that her hips were what the French would call a little too *prononcée*, a failing to which Harbinger's stock were rather prone, she was the perfection of proportion. The Rev. J. Bolden, one of the best judges of shorthorns in the county of Lancaster, offered for her, when a heifer, a blank cheque to be filled up by Mr. Booth with any amount he thought proper. Though Red Rose never aspired to personal distinction, she has not gone without her fame, having given birth to the peerless Queens—Queen of the May, Queen Mab, Queen of the Vale, and Queen of the Ocean—all by Crown Prince—four sisters more remarkable than any one cow has ever before produced—and also to Lord of the Valley by Crown Prince (the sire of almost all the young stock now at Warlaby), Lord of the Hills, Ravenspur, and another young bull by Sir Samuel.

Some space must here be devoted to the description of the four royal daughters of Red Rose, and first a word as to their names. To judge from the ludicrously unmeaning and inelegant appellations which offend one's eye on almost every page of the Herd Book, few people would appear sufficiently to appreciate the importance of a good name, which, to be good, should be either appropriate and euphonious in itself, or suggestive of pleasant associations. Nothing can be more graceful than such titles as Queen of the May, Queen Mab, Queens of the Vale, of the Ocean, and the Forest, Lord of the Hills, Mountain Chief, &c. The very mention of them calls up visions of beauty; the month of Love, with all its flowers; the Fairy Queen, with all her elves, tripping o'er hill and dale, forest and mead,

"Or on the beachèd margent of the sea."

It is impossible for a man to be *prosy* in writing of the possessors of such romantic titles.

Queen of the May was in almost every respect a model of what a Shorthorn cow should be. Her loins and chine were very wide, flat, and deeply fleshed; her quarters long and level; her head sweet and feminine; her shoulders, girth, and bosom, magnificent. Her only failing point was a want of fulness in the thighs, proportionate to the even massiveness of development displayed everywhere else. During her short career—for she was permanently injured in a railway journey, being then for the first time in calf—she won six prizes at the Royal, the Yorkshire, and the county of Durham shows, being awarded at one of the latter the 100-guinea challenge cup in 1857.

It has been reported that Mr. Booth refused for Queen of the May an offer of 1500 guineas, the highest price ever bidden for a Shorthorn. The circumstances—which are given on the late Mr. R. Booth's authority—are these:—two gentlemen from America, apparently agents for an American company, came to see the herd, and when they saw Queen of the May were completely riveted by the fascination of her beauty. After dwelling for some time upon her perfections, they inquired of Mr. Booth whether he would part with her. He replied that he "would not sell her for the highest price ever given for a Shorthorn." "That, sir," said one of them, "was, I believe, 1200 guineas?" Mr. Booth answered in the affirmative. They consulted together, and asked him whether he would take 1500 guineas, which Mr. Booth declined to do, remarking that if she bred a living calf, and he had the luck to rear it, she was worth more to him to keep, and they relinquished her with regret, leaving on Mr. Booth's mind the impression that, if he had entertained the idea, even that large amount might possibly not have been their final offer.

Of Queen Mab and Queen of the Vale no better description can be written than one that was furnished to the Highland Agricultural Society's Journal after the Perth show in 1861, when Queen of the Vale was first in the cow class, and Queen Mab second. "Queen of the Vale is a cow of faultless proportions, a perfect parallelogram in form, with well-fleshed, obliquely-laid shoulders, a good head, and very sweet neck and bosom, sweeping finely into the shoulders, the points of which are completely hidden by the full neck-vein. Queen Mab is, if possible, still more remarkable than her sister for her broad, thick level loins, depth of twist, and *armful* of flank; but she is now perhaps less faultless, as her hindquarters are becoming plain, and patchy from fat. She is, however, equal, if not superior, to Queen of the Vale in her marvellous capacity of girth, fore rib, and bosom. Like her sister, she maintains her cylindrical proportions wonderfully throughout, the ribs retaining their circular form up to the shoulders, with which they blend without any depression either at the crops or behind the elbow, and from thence the fore-quarters taper beautifully to the head." Cuddy used to say, "I dinna ken whether on 'em I sud choose: Queen Mab's t' grander colour, but then Queen o' t' Vale walks wi' sic an air! There be ya thing—yan could na be far wrang whichever on 'em yan tuke;" an opinion in which the visitor would have agreed with Cuddy.

> "Yours is, he said, the nobler hue,
>   And yours the statelier mien;
> And, till a third surpasses you,
>   Let each be deemed a Queen."

Queen Mab was the winner of twelve prizes at the principal shows, amongst them the Durham County hundred-guinea challenge cup in 1859, which had been won the previous year by Nectarine Blossom, and which having been also won by Queen of the May in 1857, as we have previously seen, now became the property of Mr. Booth, and an heirloom of the House of Warlaby. Queen of the Vale was the winner of nine prizes.

It may be worthy of mention that when Queen of the Vale was shown as a yearling at the Royal exhibition at

Warwick, she obtained no notice whatever from the judges. She had not gone through the all-essential "training!"

Sister to Windsor, one of the best of the Warlaby herd, and which had shortly before been pronounced by an eminent Shorthorn breeder to be perhaps the most perfect model of Shorthorn conformation in the kingdom, was shown, together with Chastity, fresh from the pasture, and they were both unsuccessful. Happening to be at Warlaby a fortnight before the Warwick meeting, Mr. Booth showed me these cows in the pasture, pointing with pride to the extraordinary depth and firmness of the flesh on their backs and ribs, which being the product of natural food, naturally acquired by grazing, was just what a butcher likes—"solid," as Mr. Booth expressed it, "as a well-stuffed wool-pack." They were destined for Warwick.

"Oh! that this too too solid flesh would melt,"

thought I. The cows had no "*quality.*" They were signal examples of what animals of unfashionably robust constitutions are brought to by pasture grass, open air, and exercise. The lean flesh was evidently in excess of the fat; and the fat was evidently blended with the lean, instead of being all outside of it. Moreover, the fat was not of that nice soft unctuous nature which is acquired by close confinement and liberal supplies of new milk, linseed-cake, and linseed-oil, but of the firm waxy consistence, so unpleasant in venison. I felt convinced that cows in this untrained condition had no chance of favourable recognition in the show-field, and ventured to express this conviction to Mr. Booth—a conviction which was fully realized. There are those who contend that agricultural associations, professedly formed for the improvement of animals designed for the food of man, and the encouragement of tenant farmers to vie with their richer rivals in effecting this improvement, ought not to discountenance the natural and inexpensive system of feeding which induces a healthy and vigorous development of flesh, and encourage the artificial mode, which results in a diminution of flesh, and that substitution of flabby fat to which some attach the name of " quality ;" but these people appear to look at the matter in a vulgarly utilitarian point of view.

Few people are aware what this " quality" represents. It represents the liberality of the owner, the lavish expenditure of costly food upon the animal that possesses it, the overflowing pails of new milk, the superabundant supplies of cake, corn, and condiment, and the luxurious repose and warm housing and clothing it has enjoyed. Some readers will add—" it also denotes the torpor and derangement of all the animal functions which result from this liberality and indulgence—this eating of the bread of idleness in the lap of luxury." Be it so: Nature, in default of other outlets for this excess of nutrition, deposits it—no matter whether by an unnatural and morbid process or not—in the shape of fat, where it " communicates a pleasurable and delightful sensation" to judicial fingers, and valuable parings for the tallow tub. Once do away with this " quality" test, disqualify animals with soft, or what some irreverently call " flabby," handling, and you abolish the *forcing* system. For if *firm* substance were insisted on, it would be necessary to develop lean flesh, or in other words, muscle; and, to do this, the animal would require to have constant exercise, and therefore to be what *Nature* doubtless designed a beef-growing animal to be, a *grazing* animal. Under such conditions it would no longer be those animals that had cost the most money to rear and feed that would take the prizes; but such as had the greatest *natural aptitude* and capacity for healthy and ample development. The triumph of skill, which consists in "training" animals destitute of any natural disposition to acquire flesh, and in filling out their points and covering their defects by the long-continued use of rich and costly food, would fail of its reward. Then the Royal Agricultural arena would lose much influential patronage. It would be thrown open to all sorts and conditions of men. We should have farmers and men of moderate means competing with men of affluence! We should have horrid butchers speaking approvingly of a "carcase" as being "hard fat," being "grandly marbled," and "prime flavoured;" and telling us that "such like could never be too fat, for that every ounce of them would sell at top prices like a pound o' butter!" What would become of the artistic and artificial graces of Shorthorn breeding? Once admit that beauty in a Shorthorn, like beauty in architecture and in many a work of art, should be the handmaid of practical utility, and perchance the pampered herds of Crœsus may display their meretricious charms before obdurate and hostile judges—

"May I lie low before that dreadful day,
Press'd with a load of monumental clay."

Thanks to the discrimination of our judges, a "wealthy" heifer has hitherto implied as its counterpart a wealthy owner; an animal of "quality" and substance, a man of quality and substance; a well-lined hide, a well-lined purse; a quadruped rolling in fat, a biped rolling in riches; in short, it has hitherto been a matter of necessity that he "who *feeds* fat oxen should himself be fat." It is this that has maintained the exclusive and gentlemanlike character of these yearly contests; but from the moment that we disqualify "quality," we admit "the vulgar herd" in a double sense. Nor would the mischief stop here; foreigners would buy prize animals whose constitutions had not been undermined; they would have *calves* from them, and consequently become independent of us! No, let soft touch ever continue to be the touchstone of merit, and all these evils will be averted!

But to turn from this digression. Queen of the Ocean is a superb Shorthorn—a queen of cows. She is described in the Report of the Battersea Meeting, in the R.A.S.'s Journal, as "a short-legged, well-framed, useful animal, and by far the best female in the yard, with shoulders and houghs as near perfection as possible." At the Royal Agricultural Society's Meeting at Battersea, in 1862, she won the first prize and gold medal in the cow class, and at the County of Durham Show, in the same year, the hundred-guinea cup. At the Royal Society's Meeting at Worcester, in 1863, she gained the first prize jointly with Soldier's Bride for the best pair of cows. At this meeting the whole of the five entries from Warlaby (four of which were pairs) were prize-takers. Queen of the Ocean was not forced for show until the

February before the Battersea Meeting: but ran out at pasture with the other milk cows. She has gained, in all, ten prizes. Her young bull Prince of Battersea, when only a yearling, had won splendid credentials—no less than seven first prizes, when the heat and excitement of the Royal Show Yard at Newcastle brought on inflammation of the lungs of which he died. Mr. Booth had refused an offer of 800 guineas for him, and 300 guineas for his yearly rent.

Queen of the May the 2nd, the daughter of Queen of the Vale, by Sir Samuel or Windsor, next claims the tribute of our admiration as a nonpareil of beauty. She was a heifer of faultless symmetry, and gained no fewer than fourteen prizes, including a first and second at the Royal Show at Battersea and Worcester.

Another branch of the Halnabys springs from the Sister of White Strawberry—Strawberry 3rd, by Young Matchem, himself of the Halnaby family. From her descended Bagatelle by Buckingham, a fine large cow, very wide in the floor of the chest, and with capital thighs, back, and loins. Her only weak point was a rather too upright shoulder; she was not however quite so elegant in form as Isabella Buckingham, to whom she stood second at the Exeter Royal in 1850, and at the Leeds Yorkshire Show in 1849. Bagatelle produced five calves—White Knight, presented to Mr. John Booth; Warrior, sold to go to America; Butterfly, and Benevolence, and Bianca by Leonard. Bianca gave birth to Bridesmaid by Harbinger, Bride Elect by Vanguard, and Prince of Warlaby by Crown Prince, a bull which in Mr. Booth's opinion possessed more of the character of Crown Prince than any bull he had bred. Prince of Warlaby was for many years on hire in Ireland, where he became the sire of innumerable prize animals. The cow Bridesmaid, of which an admirable portrait hangs over the sideboard at Warlaby, was an animal of deep, circular, and beautifully symmetrical frame, with a wide-spread back and loin, long, well-filled quarters, and magnificent bosom. She won nine prizes, including one second and two first prizes at the Royal Meetings at Lewes, Gloucester, and Carlisle.

Bride Elect was remarkable for the extraordinary development of her fore-quarters, and particularly of her breast, the depth and massiveness of which so far exceeded that of any Shorthorn hitherto known, as to have passed into a proverb. She was in all other respects an excellent animal, with beautiful head and horns, and admirable quality of flesh: but I shall call upon Cuddy again, to describe his old favourite; for I have almost exhausted my descriptive powers; and, indeed, when you have once done justice to the points and proportions of a single average Warlaby Shorthorn, it is but as a twice-told tale to describe others of the sisterhood; for it is the peculiarity of these tribes, and their distinguishing merit, that they are all cast in the same mould. With shades of difference and gradations of excellence that suffice for the charm of variety, their conformity in all important points to one standard is so remarkable that it may be truly said of them *ex und disce omnes.* "Aye! yon's poor ould Bride Elect. Did ter ivver see sic an a breast and sic leeght timbers? Yan wad wonder how sic lile bane could hug sae mickle beef. Look at her rumps and thighs, and loins, and aboon au, that breast. Why, there be amaist plenty for twa beasts." Bride Elect was the winner of six prizes. Her calves were three heifers—Royal Bride and Bridal Wreath by Crown Prince, and Bridal Robe by Lord of the Valley; and two bulls—Royal Brideman by Crown Prince, and a roan bull by Lord of the Valley. The former was found hanged at his stake on the eve of starting for his first year's service. He had been let for 200 guineas a year. Royal Bride was a beautiful heifer, with all the substance and shapeliness of her dam. She unfortunately took cold, which brought on inflammation in the feet, from which she never recovered. She left behind her one daughter, Royal Bridesmaid by Prince Alfred, a lovely white heifer, with her grandam's massive bosom and girth. Bridal Wreath, a beautiful young cow, of a rich creamy white colour, also inherited her dam's fine ribs and deep capacious fore-quarters, and was what Cuddy called "a rale cloggy beast." There were good judges who considered her the best cow for breeding purposes then in the Warlaby herd, though she had never been under the shelter of a roof until she gave birth to the now well known British Flag by Lord of the Valley. She afterwards bred another very promising roan bull calf by the same sire. Bridal Robe, the third daughter of Bride Elect, was a shapely heifer of that rich purple and primrose colour so pleasing to the eye.

The next tribe of Warlaby cattle whose history I shall endeavour to trace is that of the Broughtons. I have mentioned Broughton by Jerry as one of the small herd with which Mr. R. Booth, on removing to his late father's residence at Warlaby, recommenced the breeding of shorthorns. She was a cow of a superior stamp, and more than average milking capacity, and was highly thought of by Mr. Booth. She was the dam of Raspberry by young Red Rover, a very fine bull, though rather larger than Mr. Booth approved. He was accidentally hanged by getting too far back in his stall. Broughton also gave birth to Young Broughton by Young Matchem, and Lady Stanley by Lord Stanley.

Young Broughton's daughter, Bliss by Leonard, who gave rise to the celebrated family since known by her name, was a neat, medium-sized cow, of a good roan colour and with good hair. Being a very heavy milker and regular breeder, she was wont to get low in condition; but when dry, like all of her family, got rapidly into good case, and looked very attractive in her holiday trim. She gave birth to Blithe by Hopewell, Bonnet by Buckingham, and Bridget by Baron Warlaby.

Blithe, whose descendants continue the family at Warlaby, was a very neat little roan cow, with remarkably well-sprung ribs; like her dam she was a great milker, and when dry, which was seldom the case, got quickly fat. She has been known to produce two or three calves in as many successive years without ever ceasing to yield her daily supply of milk. She was the dam of an excellent bull, Valasco by Crown Prince: an indifferent one, Knight of Warlaby by Windsor; and a pretty little white cow, Lady Blithe, by the same sire. Lady Blithe has been very prolific; when only seven years old, she had produced six calves, and all of them heifers. Of these, Lady Mirth by Sir Samuel is a good thick cow, with excellent rib and loin, and great girth; Lady Joyful by Lord of the Valley (one of the pair of prize heifers at Worcester) is a massive, compact animal, full of hair; and Lady Blithesome, her own sister, is equally promising. Lady Blithe produced twin roan heifers by Lord of the Valley, whose propensity to beget twins is remarkable. The remainder of the Bliss family descended from Blithe's other daughters—Bonnet by Buckingham, and Bridget by Baron Warlaby—chiefly belong either to Mr. Peel or to the Stackhouse Herd. The few which do not are the property of Lady Pigot.

Bonnet by Buckingham was a compact, well-framed cow, with beautifully-laid shoulders. Though a great milker, she had plenty of substance, and when sold at the Killerby sale had a remarkably fine coat of long silky hair. She had been presented by Mr. Richard Booth to his brother, being at the time in calf to Royal Buck. The offspring was Wide Awake who was therefore a grandaughter of Buckingham on both sides and had Leonard as his great grandsire on both sides also, and who inherited and transmitted their perfections. Wide Awake was purchased at the Killerby sale in 1852, being then a yearling heifer, by Mr. R. Emmerson of Eryholme, and bred him three calves, one of which, Lady Grandison, afterwards became the property of Lady Pigot. From Eryholme she was transferred to Stackhouse, where she was for ten years well known as the head of its proudest family and the dam of the two beautiful cows Lady of the Valley by Lord of the Valley, and Windsor's Queen by Windsor. Unfortunately the first of them produced only bulls, so that her female descendants find their sole representative in Windsor's Queen; but the sons and the grandsons of Wide Awake have been purchased at high prices and used in many a first class herd, and when used, unfailingly appreciated.

Bonnet herself was purchased at the Killerby sale by Mr. Anderson, by whom she was sold to Mr. Wood of Castle Grove, in whose judicious hands she was destined to give birth to other offspring of equally distinguished merit. There she bred Prince Patrick by Prince Arthur, the property of those spirited breeders the Messrs. Atkinson of Peepy; but, above all, she produced, at Castle Grove, Bustle by Valiant (10989). Mr. Wood more than once refused 500 gs. for Bustle, and finally sold her to Lady Pigot, at 8 years old, for 450 gs.; and she has since bred a heifer, Bellona, to Sir Roger. Bustle produced in Mr. Wood's hands Princess Royal, Princess Maude, and Princess Helena, all by Prince Arthur, and Princess Louise by King Arthur. Princess Royal, for whom Mr. Wood refused an offer of 400 gs., died of apoplexy a month before calving; but Mr. Wood had, in 1861, the satisfaction of disposing of Bustle's three remaining Princess daughters, with her granddaughter Belle Etoile by King Arthur and from Princess Maude, to Mr. Carr for the sum of £1200, two of them being calves. Lady Pigot became the possessor of Belle Etoile, Mr. Peel of Princess Maude. The Stackhouse herd retained Princesses Helena and Louise. With the exception of Princess Louise, who has given birth to two very good heifers—Princess Beatrice (now the property of Lady Pigot) and Princess of the Blood—these fine and valuable cows have only produced bull calves. Princess Helena was obliged to be slaughtered, but she had previously bred Prince of the Purple by War Eagle, which was sold to Mr. M'Dougall, of Australia; and Princess Maude, a most excellent cow, presented Mr. Peel with four bulls, of which her twins, Hengist and Horsa by War Eagle (the former of them the second, in the two year old class at the Royal show at Worcester) and Abbot of Knowlmere by Monk, a young bull displaying very remarkable excellences, are well known. Unfortunately Lady Pigot's Belle Etoile has hitherto followed the same course, and added to the Branches Park herd members of the male sex only.

In speaking of the value of the Booth blood, it seems proper to dwell for an instant on the extraordinary prices realised by this one family. Assuming the value of Princess Maude to have been the 400 gs. which Mr. Wood had refused to take for her, the price of Bustle, her four daughters, and her grandaughter—six in all, two being calves—was £2092. 10s, or an average of £348. 15s, an amply sufficient proof of the high value of the Booth blood. Nor is this all. Offers equally large have been made for the females of the Wide Awake branch of this family. One such is thus alluded to in Bell's Weekly Messenger of October, 1865, under the head of *Shorthorn Intelligence* :—" For Lady of the Valley, when a yearling, Mr. Carr declined an offer of 400 gs.; a similar sum of money for Wide Awake when nine years old; and 250 for Windsor's Queen, Wide Awake's daughter by Windsor, when a monthling: no less than 1050 gs. for three animals. These we know were *bond fide* offers."

Of the animals which were thus valued by the public no more need be said. I may here, however, mention that Mr. Booth, shortly before he died refused an offer of £15,000 for his herd, then consisting of *some thirty animals.*

But to return to the remaining daughters of Bliss — Bridget by Baron. Warlaby has been already mentioned as one of the animals sold to Mr. S. E. Bolden. She was in calf by Crown Prince, and produced one of the loveliest shorthorns in the world in Mr. Bolden's well-known Bridecake, and afterwards two good cows, Mr. Torr's Britannia and Mr. Peel's Blissful, both by Grand Duke. Bridecake was the dam of Mr. Peel's Bride by Duke of Bolton, a short-legged, deep-framed, wide-backed cow; and Blissful was the ancestress of the rest of the Knowlmere Bliss family, amongst which is a cow by Monk, "Boundless" by name, of which it is not too much to say that she almost rivals her queenly relative Bridecake. And indeed it may be observed generally of the Knowlmere herd that besides containing Bridecake's daughter and grandaughter, Bride, and Bride of the Isles, which are nearly pure Booth, it contains also several females descended from Bliss's daughter Bridget, which have been crossed by the first Bates bulls; and to those who admire the combination of the Booth and the Bates blood it may be said that nowhere can that combination be seen in greater perfection than at Knowlmere, while Mr. Peel's judicious skill in using to cows so bred such bulls as Sir Samuel and Sir James has been attended with the best results.

Broughton by Jerry was also the dam of Lady Stanley by Lord Stanley. Lady Stanley produced Silk, which was sold to Mr. Pollock, of Mountainstown, and Satin, both by Buckingham, and was herself sold to Mr. Bolden, of Hyning.

Satin was all a dairyman could desire. In the full flush of her milk she was wont to suckle two calves and require milking dry after them,

"Bis venit ad mulctram, binos alit ubere fetus."

Mr. Booth was recommended to enter her in the dairy cow class at the Royal Exhibition; but the great quantities of rich milk which she yielded so absorbed her fattening properties that she was seldom what Cuddy calls "menceful" enough for show. She produced but one daughter, Sarcenet, who leaves no offspring, and seven bulls, amongst which was—besides Windsor 2nd, and Messrs. Booth's very good and very useful animal Knight of Windsor—Mr. Housman's admirable bull Duke of Buckingham by Crown Prince. This grand and massive animal, who, at one time weighed 25 cwt. live weight, was the only instance ever known in the annals of the Warlaby reign of a Crown Prince bull becoming the property of a subject. He was presented by Mr. Booth to his nephews at Killerby, who sold him to Mr. Housman. Many a neighbouring herd bears the impress of his worth: his son Lord of the Harem has spread wide his name; while at Lune Bank many of his valuable descendants, and particularly I may name his beautiful daughter, Queen of the Harem, still remain. His own brother, General Havelock, whom Mr. Booth retained, proved himself also a most excellent sire when let to Mr. Sanday.

The Princess Elizabeth tribe, the original progenitress

of which was bought by the late Mr. Booth of a dairy-farmer at Ainderby, has now no representatives. Princess Elizabeth by Crown Prince, was a large, lengthy roan cow, not quite so short-legged or thick-set as some of her companions, but combining milking and grazing qualities in a very unusual degree. Her own sister, Princess Mary, a compact, heavily-fleshed heifer, proved sterile; as did also Princess Elizabeth's daughter, the victorious Queen of the Isles. The latter carried all before her as a yearling in 1858, being first at the Royal Meeting at Chester, where she outshone all the gems in the Towneley diadem; first at the Yorkshire Meeting at Northallerton, where she also won the special prize of £20, and first at the county of Durham Show at Sunderland.

Christon, another of the animals with which Mr. R. Booth commenced his second herd, was from an Ainderby cow by Mr. Thomas Booth's bull Jerry, of the Lady Betty tribe. Christon herself was by Priam, of the Halnaby tribe; her daughter by Roseberry, of the Blossom tribe; and her great grandaughter was Caroline by Fitz-Leonard, of the Isabella tribe. Caroline had twelve calves. Her first were twins, calved when she was very young, and with such difficulty as to permanently distort her spine. She was a prodigious milker, giving, it is credibly affirmed, when in the prime of her life, four average pailfuls of milk in the day. These two circumstances combined to make her a plain cow, but the merits of her daughter Alfreda by Prince Alfred, an excellent, round-ribbed, strong-backed, massive white cow with a true Leonard head, and an apparently very robust constitution, indicate what her dam, with her beautiful shoulders and rich hair and colour, might have been but for her accident, and her too copious contributions to the milk-pail. Constance by Lord of the Valley, a granddaughter of Caroline, has also produced twins.

Perhaps none of the Warlaby families has been more illustrious at home and more renowned in the field than the ancient family of the Blossoms. We find Blossom, daughter of Young Red Rover, and seventh in descent from Blossom of Fairholm, settled with other distinguished descendants of Hubback, on the lands of Warlaby, in 1835. She had issue Roseberry by Raspberry, Hawthorn Blossom by Leonard, and Baron Warlaby and Cherry Blossom, both by Buckingham.

Cherry Blossom was of a fine blood-red colour, with a little white. She was a noble animal, with massive forequarters, and of stately presence. She gained four first prizes at the Royal and Yorkshire Shows, vanquishing at one of them Mrs. Mason Hopper's all-conquering Violet, for which 350 guineas had been given at Mr. Carruthers' sale.

Baron Warlaby was a rather small, but very handsome, well-proportioned bull, and a most impressive sire. He was out on hire for nine years, several of which he spent in Ireland, where he well upheld the honour of his family. In Mr. Dudding's herd he did such wonders that there was quite a run upon the Panton heifers, some of which might have challenged comparison with any shorthorns of the day.

Hawthorn Blossom was a fine, level, white cow, with great milking capacity. She had eight calves—Benedict,

Bloom, and Plum Blossom, by Buckingham; Rose Blossom, by Royal Buck; Thornberry and Highthorne, by Hopewell; Orange Blossom, by Vanguard; and Nectarine Blossom by Crown Prince.

Bloom was a large cow, of an evidently robust constitution; she had not the refinement of form which distinguished her sister Plum Blossom, but was withal a very noble cow, and had a fine udder. She was never exhibited except as a calf, at the Yorkshire Show, in 1846, where she won the first prize. She was a gift from Mr. R. Booth to the late Mr. John Booth, and gave birth at Killerby to Venus Victrix (who, with her progeny, has already been described in the account of the Killerby herd), and to four bulls, of which two, Neptune by Water King, and Dr. Buckingham by Hopewell, eventually found their way to America. Dr. Buckingham was bred by Mr. Ambler, who had purchased Bloom at the Killerby sale, by the advice of his bailiff, Mr. Dodds, a very superior judge of shorthorns, especially, where superiority of judgment is chiefly shown, in estimating the merits and future capabilities of an animal in a lean state. Dr. Buckingham was used for some time in the Sittyton herd, and from thence transferred to that of Mr. Alexander, of Kentucky. Bloom unfortunately broke her thigh in turning out of the cow-house in a hard frost, and was slaughtered when five months in calf by Grand Duke.

Rose Blossom, by Royal Buck, was a first rate animal; her loin, ribs, and chine were perfect, her shoulders neat and well out, but her hips were rather high and round, giving her very broad back a slightly hollow appearance along its otherwise level top. She won four prizes, including a first and second at the Royal Meetings in 1852 and 1853. She left no offspring.

Orange Blossom, by Vanguard, a rich creamy white cow, was perhaps even superior to her sister, but also proved barren. No one could see her without regretting that Blossom so fair should fail to fructify.

Plum Blossom by Buckingham was a level, lengthy, short-legged cow, of great substance. She had abundance of hair, of a rich purple roan, a very sweet head, and high-bred appearance. While still but a slip of a heifer (for Plum Blossom was no hot-house nursling, but a wilding of the fields from her birth), Mr. Eastwood, visiting Warlaby with the late Mr. Booth, had the sagacity to foresee the perfection to which she would mature. He made tempting overtures to Mr. R. Booth to compass her transfer to Towneley, which he flattered himself the latter did not seem disinclined to entertain; but on reviving the subject after dinner, Mr. Booth dashed his hopes, by intimating that he could not allow him to "put in his thumb and pull out this *plum*." Plum Blossom was the first prize cow at the Royal Meeting at Windsor, in 1851, and the second at the Yorkshire Show at Burlington the same year, her dam's sister Cherry Blossom being first. Plum Blossom was the parent of Peach Blossom by Water King, and of Windsor, and Own Sister to Windsor, by Crown Prince.

Peach Blossom was a handsome heifer, whose only faults were that her back rather wanted width, and her tail stood up a little higher than is consistent with neatness. She was only second to Bridesmaid in the two-year-old class at the Royal Show at Gloucester in 1852. She never bred.

Windsor—whose portrait forms the frontispiece of this work—may be said to be the Comet of modern times; he was a very symmetrical animal, of extraordinary length, with a good masculine head and horn, a well formed neck, a very deep and prominent breast, and well covered obliquely-laid shoulders; his back was admirably formed—firm and level—and his ribs were finely arched up to the shoulders, forming a cylindrical shape throughout; his quarters were very long and flat, his thighs, flank, and twist remarkably deep and full, and his legs short, and fine below the knee. From the top of his shoulder to the tip of his brisket he measured no less than 4 ft. 10 in. In the report of the Mark Lane Express upon the Royal Agricultural Meeting at Carlisle, whither Windsor had been sent straight from the pasture field, the following pertinent remarks upon him occur: "The first prize bull is worthy of special commendation, and this not only for his real merit in form and touch, his extraordinary length, that long, low, and even look, which argues so much for perfection of form; it is not only for this we would uphold him, but perhaps even more so for the condition in which—to borrow from another pursuit a most significant expression—he was brought to the post. Of all the bulls entered at Carlisle, Mr. Booth's white 'Windsor' was not only the best for shape and symmetry, but he was the best fitted to breed from. Compared, indeed, with some of the over-fed animals which stood near him, the superficial observer might wonder how he came to be placed first. It is, however, only the superficial that can be deceived in this way, while it is a very great fact to establish that a lean and really used bull did beat on his innate merit all that pampering and over-feeding could make up to show against him. As was well said by those who knew him best, he was too good for that." The writer of this report had an evidently just appreciation of the true build of a shorthorn bull. It was this "long, low, and even look, which argues so much for perfection of form," that made Windsor's *tout ensemble* so complete; with such symmetry and utility of form he could not but come with honour out of every contest; and, indeed, his motto might have been, "*Veni, Vidi, Vici,*" for though he entered the lists but ten times (at the National and Northern county shows), he won, besides other trophies, nine first prizes, and one second, being, to the surprise of many good judges, placed second to Lord Spencer's Vatican at Lincoln. After the Royal Agricultural Meeting at Carlisle, Mr. Booth refused an offer of 1000 gs. for him from an Australian breeder, who subsequently raised his bid to 1100 gs. A very beautiful, and symmetrical cow was the "Own Sister to Windsor," with her brother's splendid forehand, finely curved ribs, and firm loin. She unfortunately bred but one calf—the bull First Fruits by Sir Samuel, who, though not himself a handsome animal, has left excellent stock in the herds of Messrs. Willis, Mitchell, Stewart, and others.

Nectarine Blossom by Crown Prince was another admirable specimen of the Blossoms, the heroine of many a field day. She won five first prizes at the Royal, the Yorkshire, the North Lancashire, and the County of Durham

Shows. At the latter, in 1858, she carried off the 100-guinea challenge cup, in the cow class. The following year she was specially entered as extra stock and in-calf to compete for the cup again, a course which the words of the programme " for the best breeding animal in the yard" appeared in no way to prohibit. No objection had been made to the entry, and she was allowed to go into the ring with the rest before a word was said against it; and then proceedings were stopped, a special committee called, and after half-an-hour's deliberation, it was decided that Nectarine Blossom having won the cup the previous year was not entitled to compete again, and she was dismissed from the arena. Never since the news of poor Queen Caroline's dismissal from the doors of Westminster Abbey on presenting herself for coronation with our most religious and gracious king George IV. had popular feeling so unmistakeably manifested itself in the good town of Hartlepool; but the ferment suddenly subsided when the white rosette was seen gleaming on the frontlet of Queen Mab. Three lusty cheers rang out for the Fairy Queen, and one more for Warlaby and its victorious lord. Cuddy's beams, of which he had been so lately shorn, shone out again as he exultingly reiterated his accustomed vaunt: "They canna come ower me in a just cause." Nectarine Blossom gave birth to Fitzclarence by Clarence, a massive, short-legged, useful bull; Sir James by Sir Samuel, a remarkably handsome, robust, substantial animal, and excellent sire, as the herds of Mr. Peel, Mr. Pawlett, and Mr. Wood of Castle Grove attest; and Sir Robert, also by Sir Samuel, who dislocated his shoulder, and was slaughtered.

It is to be regretted that so many of this valuable tribe were dispersed at the Studley sale, to be frittered away or lost sight of through incongruous crossing. The Warlaby Branch appears to have had a most destructive propensity to breed bulls. Though the tribe has produced seventeen males (all of which, with the exception of Sir Robert, have been used in different herds), the parent stem from whence they sprung has broken down, and I have not unfrequently heard the fact of the exhaustion of this and the Charity tribe adduced to point a moral against in-and-in breeding, and to support the idea that fresh blood is required by the Warlaby herd. The illustration, however, would seem rather to warrant the opposite conclusion, for Charity was a daughter of Buckingham (half Craddock's blood), and grandaughter of Leonard (half Raine's); and of the three daughters of Hawthorn Blossom which failed to breed, Rose Blossom and Orange Blossom were by sons of Buckingham and their dams by Leonard, and Peach Blossom was by Water King (half Bates), and her dam and grandam by Buckingham and Leonard.

I believe that where infertility has occurred in the Warlaby herd it can in no instance be justly attributed to in-and-in breeding, but generally to causes which have led to the decline and extinction of States as well as of Shorthorn tribes.

Those causes I believe to be luxury and indolence. It will be observed that nearly all of this Blossom family have been more or less "trained" for exhibition, and have, therefore, necessarily been subjected to that system of forcing, which, by concentrating the vital energy and circulation round the digestive organs, deprives the generative ones of their due share of those important principles, and eventually the animal system of the power of reproduction.

In the human family the proportionate infecundity of the wealthy and the labouring classes is said to be as six to one, the advantage which the latter have over the former being doubtless attributable to their more frugal diet and active out-of-door life. It is fair to infer from analogy that the same causes produce the same effect in cattle. In both alike, too much and too rich food and too much ease sap the constitution, by inflaming and vitiating the blood, and impeding, by inaction, the process necessary for its purification. Thus the interior organs, which are dependent for their healthy action upon wholesome supplies of the vital fluid, assume a morbid condition, and become incapable of adequately performing their functions. It appears to have been wisely provided by Nature that the organs concerned in reproduction should be the first to refuse their office under this state of things; for where the various vital organs of the dam have been impaired, the offspring (which, as is generally supposed derives these portions of its structure from her) could not inherit perfect ones, and therefore a sound constitution. In confirmation of the preceding remarks I need only point to the fact that the greater proportion of prize animals do not breed at all, and the remainder rarely more than once or twice, whilst the offspring of the latter seldom attain to the dignity of prize animals themselves or can boast of average fruitfulness.

I would not, however, be understood to imply that the offspring of prize animals must necessarily be of enfeebled constitution. There may be, and doubtless are, animals so robustly organised that they may be subjected to this unnatural treatment without suffering such constitutional injury as to incapacite them for the production of healthy offspring; but I believe that these are exceptions which rarely occur, except, perhaps, in the case of cows of such naturally-hardy constitutions as were the daughters of Hawthorn Blossom, some of whom have been seen camping out, day and night, when the surrounding hawthorns bore no other blossom than the snow-wreath. The mischief appears to lie less in the amount than in the *kind* of condition required by the judges, a condition which can only be acquired by an unnatural process. "Ask now the beasts, and they shall teach thee." Nature has provided that the cow should obtain her aliment only by that exercise of her muscles which ensures circulation and health; and she appears, with this view, to have inspired the animal with a capricious taste, so that, however plentiful the herbage may be, she never takes her fill from the grass immediately around her, but keeps constantly moving, as she grazes, from place to place. Under these conditions, the muscular and nervous system is healthily developed, and the fatty matter is dispersed through the muscular tissues to aid this development. Hence, as the animal ripens, it acquires firmness, and when mature, is what is called "*hard fat*." But the judges, it is said, require *soft* fat—even in ripe-fed animals—that the skin should "handle" as though floating upon butter. The experienced herdsman knows that there is but one way of acquiring this butyraceous superstratum on a healthy animal, and that is by depriving her of the exercise which is necessary to the growth of muscle, or lean flesh, and cramming her with rich, fat-forming food. Under this system, which is called "training," almost all the fat which does not accumulate round the heart, liver and kidneys, is deposited externally, above the muscular substance, which it gradually in great part supersedes. The animal has now acquired " quality," and is competent to undergo the established ordeal by touch; and if, on being tried by the " rule of thumb" and fore-finger, she is found to have been sufficiently mollified to communicate a pleasurable sensation to the judicial feelers, she will probably become the sensation beast of the show-yard. It matters not that Science declares and Experience proves this obesity to be but the mask of disease, and the system which conduces to it most erroneous in principle and pernicious in practice : it is the policy of all aspirants to show-field renown to cater to the prevailing fashion.

> " Et omnes
> Impendunt curas denso distendere pingui."

Mr. Booth strongly deprecated the system, yet was obliged in great measure to conform to it. Though his cattle were absolutely unrivalled in their aptitude for healthy and ample development on pasture, and he repeatedly sent them in blooming health and burly case from the pasture to the show-field, and occasionally with success, he too frequently, on such occasions, came off second-best. Hence it was found necessary to subject, for an adequate period, such of them as were designed for exhibition to a system whose disastrous effects even the vigorous constitutions of his cattle were unable finally to counteract. Hence the failure of female representatives of the Blossom and Charity tribes, and that in a herd whose unforced members are so prolific that no less than six cases of twins occurred at Warlaby in the first four months of 1864, in every instance the sire being a Warlaby bull, and in four cases the dams pure-bred Shorthorns.

From a book lately published by M. Caunière, entitled "De la Medicine Naturelle, Indo-Malgache," it appears that in a well-constituted adult the proportion of fat is not more than about a twentieth of the weight of the whole body. It may exceed that proportion to a certain extent without inconvenience, but it becomes a regular disorder when it reaches one-half. If there be any analogy between the human and bovine animal economy, as is generally supposed, what must be the state of a beast in whose system (as is not unfrequently the case amongst prize cows and sheep) these proportions of flesh and fat are reversed? It is idle, however, to condemn those who adopt this ruinous system, so long as the judges award their prizes and the public their commendations to animals in this unnatural state. The breeder must follow the fashion, or be left behind in the race. As systematically as the ancient husbandman selected from his herd this to propagate the breed, and that to bleed a victim at the sacred shrine, must the modern shorthorn breeder, who would maintain his position before the world, yearly single out the choicest of his herd for immolation on the altar of the Royal Agricultural Moloch. Nevertheless, hard as it might be, with such cattle, to forego the triumphs of conquest, good sense would seem to say—if such be the cost at which renown is to be purchased—

"Then rather let my herd, as leisure leads,
　Wanton inglorious o'er the grassy meads."

But though Mr. Booth deeply regretted the necessity of showing his cattle, he still felt himself compelled to do so even to the last, and he showed Prince of Battersea, at Newcastle, only a few months before his death. Still, owing partly to his unwillingness, partly perhaps to the increasing infirmities of old age he scarcely showed so many of them, or showed them so extensively as he had once done. But notwithstanding this the Booth blood was fated still to maintain that proud pre-eminence which it has always held when fairly tested. Animals of kindred blood, in the language of the Turf, "took up the running." To this blood Colonel Towneley's splendid and unconquered bulls Master Butterfly and Royal Butterfly were in some measure indebted for their victorious career, the common dam of them both being by the Booth bull Jeweller, while many others of Colonel Towneley's numerous prize-winners had a still larger amount of it in their veins. Mr. Ambler's well-known success was also in a great measure due to animals wholly or principally of the Booth blood.

But Warlaby's great supporter in the last days of its late owner was Mr. Booth's fair friend and ally, Lady Pigot. A devoted and enthusiastic admirer of the Booth cattle, she purchased them at the highest cost, showed them most extensively, and conquered in a hundred fields.

To enumerate all these prize-takers would be impossible and superfluous, but one we must mention, ROSEDALE— a name which must last as long as Shorthorn records may endure; ROSEDALE, perhaps, the most beautiful heifer England had ever seen in her show-yards since the time when Queen of the May electrified them, and whose many victories recalled the memory of the proud triumphs of Necklace and of Bracelet. Descended from a celebrated Booth cow of Mr. Maynard's, and herself a daughter of Valasco, to whom her dam, Rosey, was in calf when transferred from Stackhouse to Branches Park, this lovely heifer was wholly Booth with the exception that one-eighth of her blood was derived from the never-beaten prize bull Belleville. All her victories were won before she had completed the age of two years and a half; while nothing can show more positively the strength and stamina of the Booth cattle than the circumstance that Rosedale, now the property of the Duke of Montrose, notwithstanding all her training, has since bred with the greatest regularity; while her dam, Rosey, the property of Mr. Storer of Hellidon, though now between 14 and 15 years old, continues to give birth annually to a living calf, and after her last milked well for eight months, the last five of which she was again in calf.

The farm at Warlaby lies well together, and grows large crops of wheat and beans, and fine turnips, and, since it has been drained, good oats. The land is very clean, and kept in good heart. Willie Jacques, the Foreman, " reckons nout o' your hawf muckins; it's all gane in a minute, it is. Aye, gie it planty, and ye'll git summat out. Whats t' use o' fiddlin o'er nowt?" The pasture land yields a luxuriant herbage, of mediocre quality, which appears, however, to suit cattle admirably. The usual mode of rearing the calves is by allowing them to suck either their dams or nurse-cows, giving them, in addition, after three months old, a little cake and corn. They are allowed daily exercise in fine weather, and after the first winter many of them are never housed again till near calving.

The nurse-cows were put to the high-bred bulls, and their calves were reared on porridge made of blue milk and oatmeal. The males were steered, and fed off at pasture on grass, hay, and turnips. They were generally sold at about 30 months old, at from £30 to £40. Some of the females were also fed off: others recruited the staff of nurses. Many of the half-castes had to rough it in the pastures, even through their first winter, and, indeed, not a few of the shorthorns had only another year's grace before being exposed to the like hardships. The innutritious winter herbage was occasionally eked out by a supply of turnips scattered over the pastures, and when the ground was covered with snow the cattle were supplied with hay, which was stacked in the middle of some of the fields; but in an open winter they were left to earn their own livelihood.

A visit to the old Squire of Warlaby reminded one of the visit to Sir Roger de Coverley, in the *Spectator*. His establishment in-doors and out-doors consisted chiefly "of sober and staid persons; for as he was the best master in the world, he seldom changed his servants; and as he was beloved by all about him, his servants never cared for leaving him; by this means his domestics were almost all in years, and grown old with their master." Nor did the parallel stop here; there was between Mr. Booth and the worthy vicar of his parish, Mr. Raw, the same mutual esteem for each other, and hearty concurrence in works of charity and benevolence that subsisted between the good old knight and his chaplain. But though he, alas! is gone the presiding genius of the place, the memory of many friends yet recalls the image of the good old man, and of Warlaby as it was when Mr. Richard Booth was still alive.

Why should I speak of Cuddy, that renowned custodian of the herd, "whom not to know argues oneself unknown," but to say that Cuddy at home and Cuddy in the showyard were one and the same Cuddy from top to toe? The same brown wide-awake, the same variegated kerchief knotted round his neck, the same brown "kytle," and leather gaiters, and the same bland smile of welcome and reverential Eastern salaam, met you at the fold-yard gate, as in the avenues of the show-field. No Jack-a-Dandy "airs in dress or gait" ever took possession of this bronzed and horny-handed receiver of first-class medals. "Wad it beseam me to be donnin', and prenkin', and dizenin'? Is'e nobbut Cuddy, please yer honour."

If ever there was anything wrong in the showyard, Cuddy was the foremost man to lend a helping hand, and it was he who courageously grappled and held Lord Feversham's great red bull when he unfortunately killed a poor fellow at Northallerton. Mr. Raw was nomenclator of the Warlaby herd, or as Cuddy expressed it, t'parson kersen'd 'em, and it was not the least onerous part of the old herdsman's work to burden his memory "wi' sic like new-fangled names. Yan canna bethink yansel on 'em. If yan du nat bottle 'em up vara tight they gang clean out o' yan's mind. It's a wonder where foaks rakes 'em up." On one occasion Mr. Booth had been so long confined to the house that one or two of the calves had grown quite out of his remembrance. "Cuddy," said he, "what is this calf?" Cuddy tucked his whip under his arm that he might be free to scratch his "pow," and grasp at any mote of recollection that might float before his mental vision, and after standing for some seconds in the attitude of a meditative jackdaw, thus delivered himself: "Weel, sir, I canna reeghtly think on of it names, but it's t'calf as were calven t'neeght as ye runned for t'bellowses." I need not say that this little incident proved a sufficient remembrancer to Mr. Booth. Cuddy had been Mr. Booth's herdsman forty years, and his wife had assisted him in that capacity about twelve.

Mrs. Cuddy, the foddering and littering wife, a small spare form well stricken in years, was always to be seen briskly stepping about, broom in hand, now sweeping the causeways or cow-byres, now distributing "skeps" of cut turnips to the inmates of the various loose boxes, now bobbing her quick curtsey to the visitor. A very pattern to her sex was the industrious dame in her severe discountenance of all idle gossiping. Her voice was seldom heard except in an occasional snap at her lord and master, Cuddy, when all was not to her mind, or her spouse had paused from his work to have a crack aboot the beeas. Nanny—for such was Mrs. Cuddy's christian name—was acknowledged to be "a Britoner for bravery." "Tak care," she would say to a visitor going up to Crown Prince, "he is not to depend on," and she would step up to him with a basket of turnips, and rub his old head confidingly. "Hearing," says a visitor, "a cry of 'Cuddy! Cuddy! come here! here's Prince o' t'Isles loose,' I went to the yard, and there was Nanny waving her broom about, and keeping the huge animal at bay. 'I wad hae teed him up,' said she, apologetically, 'but he is sae high I canna reach his chain round his neck to fasten him.'"

Cuddy had a valuable coadjutor in that ubiquitous factotum, trusty John White, who was butler, waiting-servant and valet to Mr. Booth, and registrar-general of the births, deaths, and marriages, and all else that transpired in the Warlaby herd. John's father had been thirty years on the farm, and he had himself lived there in various capacities from his boyhood, and remembered the jubilation occasioned by the birth of Monica's twin daughters—Medora and Modesty. When illness had confined Mr. Booth to the house, and Cuddy had become less active, John made it his business, in addition to his household duties, to keep a watchful eye on the cattle—especially the young or ailing ones—in the neighbourhood of the house. So admirably did he discharge this self-imposed duty, so methodical were his habits, so retentive his memory, and so scrupulous his observance of his master's orders, that the *active* management of the herd mainly devolved upon his shoulders, and Mr. Booth found him an invaluable auxiliary.

Last, not least, came doughty Willie Jacques, the farm-bailiff, who had been upwards of forty years in the family. He first lived with Mr. R. Booth at Studley, who sent him to Warlaby in the old master's time, to take the management of the arable land and work-people. Willie Jacques's pride was rather in the nameless nondescripts of the farm, the bullocks and half-bred heifers, which converted his marvellous root and clover crops into goodly rounds and lordly barons of marbled beef, than in the pampered aristocrats of the herd, born to consume the fruits of the soil whether earned or not. Proud as Willie was of their triumphs in the show-field, nothing exasperated him like the failure on the part of any of them to contribute their yearly quota towards the increase of the herd. Willie Jacques had a capital head for tillage and general farming, and was always at his post, from which nothing could move him but the Christmas Fat Show at Smithfield. "I'se seea thrang I canna gang," was his answer to all other invitations. Curt of speech and unceremonious in bearing was Willie Jacques in his sturdy northern independence; but get him upon the subject of his kind old master, and all the frost of his nature melted away, and you found that under that dry, almost blunt manner, a heart as kindly as a child's was hidden. In one of the rooms at Warlaby hung an admirable portrait of this highly-respectable and respected steward of the Warlaby estate.

But there was one other personage, to forget whom in a sketch of Warlaby would be fatal to the character of any historian—a personage who, though seldom visible, has contributed to the visitor, perhaps not the least comfortable reminiscence which an Englishman carries away with him from any place of passing interest; and that is Ann, faithful Ann, that white-bibbed paragon of natty spruceness—the housekeeper. She came nobody knows how many years ago, to nurse the former housekeeper, an old friend of hers, who was ill, and who died at Warlaby; and Ann continued until Master could find one to suit him, which he never did, and so Ann remained still; and many are the visitors who can testify to the excellence of the pigeon pies, apricot tarts, and other delectable cates, which those brisk and clever hands have fabricated.

If I have not been deterred from attempting this simple portraiture of rural life by any alarm, lest

> "Grandeur hear with a disdainful smile
> The short and simple annals of the poor,"

it is because in the most imperfect sketch of their homely joys and hopes and fears, there are touches of nature that prove how near is their kinship to the possessor of the most illustrious pedigree in Burke's Human Herd Book, and how false is the philosophy that would ignore it; and also because it is certain that much of the happiness of Mr. Booth's useful, though unostentatious life, was reflected from the happy faces around him; much of his prosperity from the feeling in which he and they alike participated—that they were all, master and servants, members of one body, with a common weal, a common interest; and that in caring for the well-being of each other they were caring for themselves—a truth on the practical recognition of which the happiness and prosperity of agriculturists in general, doubtless, in great measure depends. The grand principle which has been preached for 1800 years, and of which Thackeray and Garibaldi have been the latest and most eloquent exponents—that true nobility has no necessary connexion with noble blood, but with honesty, gentleness, self-denial, and courageous endurance of hardship—is beginning to come home to us at last; and when we consider how faithful to the trust reposed in them are the majority of these sons and daughters of toil, how respectful to their masters, and generally considerate of the feelings of each other, how unselfishly helpful of their neighbours in want and sickness, and how heroically patient and cheerful under their own many privations, we cannot but admit the truth of the doctrine, that there may be true nobility at the plough, and in the cottage.

> "Honour and shame from no condition rise;
> Act well your part; there all the honour lies."

The scene I have described, and many of its features are no more. The good old man,* of whom it may be said most justly

> "Multis ille bonis
> Flebilis occidit,"

has met with the resignation of a Christian that last summons "for which," as was said at the time in one of his obituaries, "his whole life had been one long preparation." He sleeps in peace beneath the shade of the old grey tower of Ainderby, which looks down upon the scene of his useful and quiet labours. But Warlaby is there still, and his kith and kin retain its hall and herd.

* Mr. Richard Booth died on the 31st Oct. 1864, at the age of 76 years.

And it may be added—for it is a circumstance too well-known to savour at all of flattery—that his nephew and successor, Mr. T. C. Booth, is no unworthy or unskilful heir; while his amiable wife lends a new charm to the old place; and his rising family gives the promise of the continuance of the long-continued Warlaby herd for generations yet to come.

But to return: Before hazarding any remarks of my own upon the principles and practice of breeding pursued by Mr. Booth, I propose to introduce some remarks upon that subject, written by Mr. Wood, of Castlegrove, a gentleman whose excellent and well-known herd, bred for a length of time under his own careful superintendence, adds ten-fold weight to his long-weighed opinions.

"*Castle Grove, March* 29, 1864.

" The following remarks on one particular branch of the subject of shorthorn breeding are the result of many years experience, but they are suggested to me at present by some facts which have recently come under my notice. It must frequently have been observed that animals sold at shows and at Shorthorn sales, though good in themselves, and, it may be, descended by several crosses from purely bred and perhaps well-formed shorthorns, rarely in their progeny meet the expectations of their purchasers. What is the reason of this, and why are the hopes of those persons so often disappointed? If you look into the Herd Book and examine the pedigree of these animals, I think you will almost invariably find in them recent *new* crosses—that is, recent crosses of animals of different families not related in blood. The progeny of such crosses, when good, I can never consider otherwise than as *good only by accident;* for however excellent the parents themselves may have been, I believe that the chance of their producing good animals was in proportion, not so much to their own apparent excellence, nor even to the number and qualities of their ancestors of different families, as to the number of recent good crosses they may have had of the *same* blood or family. It is this *continued relationship* in blood which gives character to a stock, and fixes its qualities, either good or bad, according to the skill and perseverance of the breeder. It would seem as if every ancestor or cross introduced more or less new elements, and that every new element multiplied immensely the possible number of new combinations in the progeny. When many of the elements introduced by a cross are similar to those already possessed by the cow or the herd, as the case may be, the chance of producing animals resembling the parents is greatly increased, and character and uniformity in the herd is obtained or approached; but where many of those new elements introduced by a cross are dissimilar to those already obtaining in the herd, the number of possible new combinations is increased, and of course the chance proportionally increased of producing animals unlike their progenitors, and unlike each other, and greater variety and uncertainty is the result; hence in herds so bred there is little uniformity of type. We sometimes see in them a good animal, very often a bad one, and we frequently see own brothers and own sisters with little or no resemblance. The principle enunciated above, if carried to an extreme, would lead to the conclusion that the closer the affinity of the animals bred from, the greater the probability of good produce, provided those put together were themselves good; so that when own brothers and sisters are both good, they ought to be put together; and if their produce should prove good, *they* also ought to be put together, *and so on*. Now, the mistake in the above deduction would seem to me in the *so on*, or in carrying the argument to extremes. In fact the practice of breeding from animals allied in blood has been followed by the Messrs. Colling and other eminent breeders, with results so satisfactory that it would seem, if not carried to an extreme, to be the best, as it certainly seems to be the most natural, course in breeding. We know that gregarious animals in a state of nature roam in flocks, to which they are very jealous of admitting strangers, so that the animals of each flock must be related in blood, probably, by many different relationships. Perhaps from this cause, as well as from similarity of habits and feeding, uniformity results, and these flocks have a fixed and steady type, and are not found to degenerate. But the case of the Chillingham wild cattle sprung from one cow and her bull calf, offers a still more striking proof that animals may be bred from near relationships for many generations without suffering any diminution of their hardiness, or of their original good qualities. The Messrs. Booth's herd has been bred for *many years* from animals related in blood by *manifold relationships*—there is no animal in their herd that is not related to each of the others in many different ways; but the practice of breeding from the closest affinities (that is from brother and sister and from parent and offspring) has not been adopted by them as a general rule—never beyond what was considered the necessity of the particular case, or as a mere experiment. If there be then any error in breeding in-and-in, as it is called, from good animals—and I confess I think there is—it can only be in carrying the practice to an extreme, or continuing to breed from the closest affinities. The laws of nature have limits which cannot be passed with impunity; what is good in certain quantity is not necessarily so in double that quantity. What is good in moderation is invariably bad in excess. The practice of breeding from animals related in blood has, in the hands of the gentlemen above named, produced the happiest results; but in less judicious hands it might no doubt be carried too far. From what has been said, another question will suggest itself. If it be true that breeding from a good sire and dam does not necessarily ensure a good progeny, can it be true that " like begets like "? I answer, that I believe that maxim to be true in a certain sense, but it undoubtedly is not true in the popular sense in which it is used, and I believe it has led many a young breeder astray, by inducing him to believe that when he had purchased a good-looking sire and dam he had secured all the necessary conditions for a good progeny. There is no more prevalent error among young breeders, and there can scarcely be a more fatal one. An animal has certain qualities apparent to the hand and eye; it also has hidden qualities that neither the hand nor eye can detect, but which hidden or latent qualities descend to the offspring, and, when the animal has been crossed with another animal of different blood, will produce new combinations palpable and unexpected. The above maxim is true then in this sense, that, though the offspring may appear unlike either parent, yet the peculiar properties of the parents are not lost in the offspring—they are inherited, but in combination may have produced effects that probably had not, and could not with any degree of certainty, have been foreseen. That these qualities are not lost would appear evident, as it is found that peculiarities of even remote ancestors will from time to time, more or less frequently, according to the skill and perseverance of the breeder, show themselves, or crop out, to use a geological expression.

"The Messrs. Booth have by long-continued and skilful selection produced good animals, and by persistently using (with occasional exceptions) animals of the same blood or family, they have obtained uniformity of type; the result is, that a bad or indifferent animal is rarely if ever produced in their herd—that is, the chance of producing one is reduced to a minimum, and so fixed are its qualities, that, to whatever part of the world members of it are removed, whether to any part of this kingdom or abroad, '*per varios casus per tot discrimina rerum,*' they invariably preserve their uniformly high character.

"I would guard against being understood to say that a cross of new blood ought never to be introduced into even old-established and good herds; but I do say that such crosses, in such herds, ought to be used rarely and with great caution, and that their use should be confined to a portion only of the herd, until the full effect can be ascertained.

"There are then, I conceive, latent qualities in animals, the effect of which in new crosses cannot be foreseen by the most skilful breeders, and the result of such crosses can only be satisfactorily known after several generations of the animals have passed. The reasons I have above assigned may be true or not; their truth does not admit of mathematical demonstration, and I do not pretend to insist on it dogmatically. I merely suggest them as possible operating causes of results that all experience proves, viz., That uniform qualities in a shorthorn herd can only be secured by breeding from animals related in blood, and high probability of excellence only by continued and skilful selection from such animals; but yet, that there is a limit in the affinity of those to be put together that it would not be safe frequently to overstep.

"That continued breeding from the closest affinities would eventually cause delicacy of constitution and diminished fertility I have good reason to believe; but I do deny that to breeding in-and-in is to be attributed the want of fruitfulness which is so generally complained of by breeders of what are called improved herds. Among the causes at work to produce that defect (for it cannot be denied that it does exist to some extent), the principal I conceive to be the three following: First, the forcing system; second, the unnatural treatment of the animals, the bulls in most cases being kept in the house all their lives; third, but not least important, is the tendency to admire and use bulls of effeminate appearance—bulls with what is called *sweet heads and horns*, that without close inspection one would mistake for steers. This is a crying evil, and the popular taste is too much encouraged by the decisions of judges in public show-yards.

"J. G. Wood."

The force of Mr. Wood's remarks on the uncertainty of breeding from *omnium gatherum* herds will be readily admitted by every scientific breeder. Unexpected varieties of form and qualities must result from a hodge-podge of heterogeneous elements, got together by miscellaneous selections from various sources. Where this diversity prevails in a herd (and in how many herds, accumulated by the most lavish expenditure, does it not prevail?) it is sufficient proof that the owner has no clear or well-defined standard of structure in his own mind to which he is aiming to assimilate his stock, and that, however he may pride himself on his independent judgment and his freedom from prejudice, he is but a novice in breeding after all.

Uniformity, or "the counterpart presentment" of the same type in every member of a herd or flock, is the true test of a breeder's skill; and in this particular the Warlaby herd bears striking testimony to the master-hand of its founder. The late Mr. Booth prided himself on having them all of one sort, and held the opinion that to have them of one sort you must breed them from one sort; and that this system is especially necessary where your object is to produce male animals that may be depended upon to transmit a definite character to their offspring. This family likeness running through the Warlaby cattle, the well-known propensity of its bulls to transmit it, and their consequent value to all breeders who wish to establish in their herds uniformity of character, has conduced to the supremacy of the Booth blood throughout this country. But if in England and Scotland the Booth cattle incontestably stand first, in Ireland they are all-in-all, as the records of almost every show-field attest, so that Ireland may be said to be virtually a Booth province. A striking example of this may be mentioned. At the Royal Irish Show, held at Kilkenny in August 1863, two splendid bulls contested for the championship of Ireland. Three eminent Englishmen were the judges, and it was long before they could arrive at a decision, so nearly balanced were their merits; yet it was not a question of rival strains of blood. Both the competitors were virtually Booth bulls; they were Soubadar by Prince of Warlaby, the property of Mr. Richardson, of Glenmore, and Paterfamilias by Lord of the Valley, then the property of Mr. Cooke, of Ballyneale House, and since of Mr. Storer, of Hellidon.

It may be worth while to remark here, as illustrative of the influence to which I have before adverted, of change of climate on the Shorthorn, that the change to Ireland has been observed by attentive breeders to produce greater development of hair; and partially the same effect is produced at home by a more natural treatment than is usually pursued—namely, allowing the young cattle, when old enough to bear it, to live summer and winter in the open field, being there supplied with such food as they require, and having the shelter of an open hovel. Such a change in some respects neutralizes in the offspring the ill-effects of that confinement and high feeding which the necessities of the show-yard too often require one or both of the parents to be subjected to, and restores again the wonted vigour.

Intimately connected with change of climate, and, indeed, inseparably united with it, is change of food; for the former necessarily involves the latter. When cattle are removed only from one county to another, or from one to another part of the same county, or even to a different farm only, supposing the mode of treatment to remain the same, yet is there necessarily a change of food. This the very pastures themselves will supply; the grass, the hay, the mangold, the turnips which their new residence produces will essentially differ from the products of the old one, and have the charm of freshness; and the consequence frequently is that under such a change cattle rapidly improve without there being any sufficient cause to account for such improvement beyond the mere change of climate, scene, and diet. Such a change appears to cause that beneficial effect in the bovine constitution which it is well known is produced upon man. Is it not well authenticated in the history of various animals, that one known and recognised aboriginal type has given rise to endless and wonderfully differing varieties, which the naturalist accounts for by the difference of food and climate to which they have been for generations subjected? And what is true of the dog, or of the sheep, is true also of the cow.

One instance which has occurred almost within the memory of man may be adduced as an example. The small polled Galloways of Scotland—which some even of ourselves in our younger days may remember travelling in large droves to the eastern counties of England, the size of them being on the average about that of a medium-sized donkey—have given rise to those large, stately, and heavy beeves which we now dignify with the name of Norfolks, and of which the eastern counties are so justly proud. Even in the individual, change of food and climate combined not unfrequently produces an extraordinary effect. Many instances might be given where, in consequence, fertility, has been renewed, and where cows which were considered hopeless have again become breeders. And no wonder; for, as an experienced breeder and highly scientific gentleman has well remarked, "Animals living year after year in the same place are eating themselves over and over again." In many respects, therefore, change of food and climate may be considered equivalent to change of blood, and the greater the change proportionably greater the effect. So that there was some scientific skill as well as some pecuniary profit in the re-importation of Duchess bulls from America, to which the admirers of the Bates blood recently resorted.

The views and practice of the Messrs. Booth appear to have been in conformity with those of all early improvers of domestic cattle. Their principal aim was to raise a *useful sort*, a sort that, besides possessing beauty of form, would milk copiously, fatten readily, and when slaughtered turn out satisfactorily to the butcher and the consumer. With this view they sought to reduce the bone of the animal, especially the length and coarseness of the leg, the prominency of the hips, the heavy bone of the shoulders, and those unsightly projections called shoulder-points. Mr. Richard Booth's preference for a moderate-sized animal was even more decided than that of his predecessors; for, apart from the difficulty he saw in obtaining neatness of form in a large animal, it was his persuasion that a moderate-sized beast was more generally useful and profitable, as yielding more milk and more beef in proportion to the food consumed by it, than one of larger size. He endeavoured, therefore, to reduce the frame into still more compact dimensions, and to improve its contour by shortening and lifting the loin, and giving it firmness and thickness, thereby further securing convexity to the back ribs, and contracting the space between them and the hips. It is this that has imparted to the Warlaby cattle that support, tenseness, and consequent straightness of the paunch, which Mr. Booth deemed all-essential to quick and economical fattening. He saw that these deviations from the then recognized standard of form would effect a reduction of the inferior meat, and augment the proportion of the more valuable joints by increasing the quantity and improving the quality of the flesh on the loins, rumps, and fore and mid ribs.

In a letter which appeared some time ago in one of the journals, under the signature of "Melibœus," occur the following remarks on the points and capabilities of the Warlaby cattle:—"A recent visit to Warlaby, with a view to hire a bull, has confirmed my own opinion, and justified the general decision of competent judges, that there is no herd in Great Britain at all fit to be compared, for breeding purposes, with that of Mr. Richard Booth. The tendency of the Warlaby animals to put on flesh *in the best places*, and to put that flesh on *quickly* with ordinary keep; their ample dimensions of structure, indicating capacity for healthy organization, and accordingly promising robust and enduring constitutions; the invariably strong deep loin, round prominent fore-rib, barrel-shaped crops, large girth, deep heavy flank, and almost perpendicular fore flank; the long level side-line below the ribs from flank to hockster or elbow; the straight under-line of the belly, the full thigh descending low down, and the twist abundantly developed; the back, straight in profile, wide and flat upon the top, and padded thickly with fine lean flesh; the quarters long, level, and well-packed; and the bosom, grandly built out, massive and symmetrical, with shoulder points buried in flesh; the short fine-boned legs, waxy horn, and thick mossy hair—these are characteristics prominently exhibited in the shorthorns of Mr. Richard Booth; and these, combined as they are with excellent milking qualities, as exemplified more or less in all the *unforced* cows, are characteristics which sufficiently declare the value of his herd. No wonder, indeed, that the bulls of Warlaby should be eagerly engaged; for the man who introduces a Booth bull among his cows is taking the best method of securing a good herd with the least loss of time, and, I may add, with the least expense too. The effect of such an introduction is immediate, and he who judiciously adventures it is surprised to find, within an incredibly brief period, animals of both sexes adorning his pastures, bearing the unmistakeable stamp of Warlaby. I have seen no bulls from any other herd which so quickly produce an improvement in the stocks among which they sojourn, and which are so frequently the sires of animals winning honours in the show-field. 'It is patent'—as Mr.

Douglas well observed in his admirable speech at the dinner given in his honour at Haddington—'that nearly three-fourths of all the prize animals at the national and other important shows are of the Booth blood, and not a few of the most successful exhibitors of the present day owe in a great measure their position and popularity to this strain.'"

As the opinions of so successful a breeder as Mr. Douglas must ever carry with them great weight, it may not be out of place here to give a further quotation from the speech above referred to. "It is not," said Mr. Douglas, "animals of a large scale that are wanted. In such subjects there is generally a preponderance of bone, long back, weak loins, flat ribs, and much coarse beef; what we want, in my opinion, is an animal of apparently small scale—but in reality not so—having a great propensity to fatten; on short legs, with fine bone, massive compact body, wide chest, ribs well sprung, thick loins, and well filled-up quarter, with deep twist, body all equally covered over with heavy flesh, and plenty of soft hair, and having no coarse beef on any part. This is my standard of a shorthorn, and when I speak of such I have in my mind's-eye many of Mr. Booth of Warlaby's best animals. Look at the docile, even intelligent expression of countenance, the waxy horn, moderately-short neck, full neck vein, prominent bosom, beautifully-laid shoulder, capacious chest, ribs well sprung from the back, thick-fleshed strong loins, deep flanks, hoggins well covered, lengthy well-packed hind-quarter, with deep twist, on straight legs, and fine bone—such are nearly all the animals that constitute Mr. Booth's celebrated tribes or families of shorthorns. There can be no mistake about the character of this herd: it is so indelibly stamped, that any person once seeing them would again detect the likeness of the herd, even in the killing-booth. In brief, I consider a perfect specimen of the shorthorn one of the most beautiful objects in creation."

Under the plastic hands of the great breeder, the Warlaby cattle have perhaps assumed as much beauty of outline as is consistent with that utility of form, at which (whilst studying to combine the two) Mr. Booth has primarily aimed. It was said by Mr. Gladstone of a man not more remarkable for the consistency and tenacity with which he applied the principles of true art to the manufacture of earthenware, than Mr. Booth has been in moulding the fickle forms of animated nature to his will, that "his most signal and characteristic merit lay in the fineness and fulness of his perception of the law which teaches us to aim, first, at giving to every object the greatest possible degree of fitness and convenience for its purpose: and, next, at making it the vehicle of the highest degree of beauty compatible with that fitness and convenience which it will bear. It does not substitute the secondary for the primary end, but recognizes as part of its business the study to harmonize the two."

Mr. R. Booth's success as a breeder was due to his clear conception and persistent adoption of this principle. Regarding economy of production, and adaptation to use, as the primary objects which a shorthorn breeder should keep in view, his exertions were mainly directed to the promotion in his herd of the tendency to early maturity and rapid fattening. His object was to produce animals with constitutions suited to the severity of our climate, and with a natural aptitude to convert that sustenance to which Nature has adapted all the organs of a grazing animal—pasture grass—into flesh of a superior quality. Complete and unrivalled as was the success of Mr. Booth's efforts in this direction, he did not achieve it at the sacrifice of Taste, but succeeded in combining with these more sterling qualifications those pleasing features and proportions of structure which an immense majority of show-yard awards has stamped as the highest existing standard of Shorthorn excellence.

The late Mr. Booth succeeded in imparting to his cattle a length of quarter such as no other herd can boast, a marvellous fulness and depth of thigh, and of the twist, or junction of the inside of the thighs, and a perfectly parallel and almost perpendicular position of the hind-legs.

It was, however, to the ample and symmetrical development of the fore-quarters that Mr. Booth's especial attention was directed. He increased the obliquity or backward inclination of the shoulder-blades, thereby preserving the level line of the back, and promoting the free and graceful carriage of the animal, and under the conviction that ample scope for the vigorous action of the heart and lungs was an essential condition to the formation of good blood, and therefore of good beef, it was his aim to improve the form and enlarge the capacity of the chest. With this view he endeavoured to augment the prominence or circularity of the fore-rib, and the width of the sternum or floor of the chest between and behind the fore-legs.

It is to the success of Mr. Booth's efforts in this direction, and the extension of surface which this improved formation of the chest affords for the accretion of flesh, that we are indebted for those valuable and almost peculiar characteristics of the Warlaby cattle—the perpendicular fore-flank, which drops even with the arm, the roundness of the barrel-shaped crops, and the width and massiveness of the projecting bosom.

To this conformation also may probably be due the very remarkable immunity from pleuro-pneumonia and other chest affections which the Warlaby and its kindred herds have hitherto enjoyed. It may here be remarked that this development of the fore-quarters was mainly effected by the free use of the male descendants of Isabella by Pilot.

The necks of the Booth shorthorns are worthy of note. Whilst displaying at their junction with the head much of that fineness and cleanness which characterise the dairy cow, they are remarkable for the bulky, yet symmetrical development into which they gradually swell as they approach and blend with the shoulders and breast, completely hiding, even in the unforced animal, the shoulder points. Ask the butcher or the grazier why he passes his hand over the base of the neck, and he will tell you that the fulness of muscle and neck vein there, afford the surest token of substance, or tendency to substance, over the whole frame.

It has been objected to the Booth cattle that their necks, though fine enough at the setting on of the head, are too short, and thicken too rapidly towards the breast and shoulder. It has been urged that they should be lighter, thinner, and more gracefully curved, which it is said would give gaiety and style to the animal; nay, I have heard an eminent shorthorn breeder assert his predilection for a bull that comes out of his box like a high-mettled horse, with arching crest, dilated eye, and featly prancing mien, instead of the solemn and laggard port which distinguishes the solid, stolid bulls of the Warlaby tribe;—a poetical idea of the "lord of lowing herds," which would be all very well if we rode our bulls to hounds, or "went in" for "glory" and "ladies' lovely glance" by encountering them in the arena, but which shows an inappreciation of the purposes for which the bovine animal was designed. What we want in a shorthorn is the reverse of all this—a placidity and composure of mind, a phlegmatic disposition, suggestive of fattening propensity.

Without subscribing to the axiom " that Beauty never deigns to dwell where Use and Aptitude are strangers," it may be safely asserted that the proportions and the disposition which best accomplish the particular purpose for which an animal was designed, are the true and desirable ones. Nature has given the horse a light elastic neck, supported by a series of small, smooth bones, springing arch-like from the shoulder, as best adapted to an animal designed for active motion and obedience to the rein. The neck of the ox, on the other hand, she has framed, with an equally wise providence, of short heavy bones, with numerous transverse processes, like the spikes of a Norwegian harrow, and, as Youatt tells us, "all these widened, roughened, tuberous bones are for the attachment of muscles, the accumulation of flesh," not of beef for Belgravia perhaps, but of wholesome food for uncritical appetites. This being so, to encourage, instead of thwarting the benevolent design of Nature, and to make two pounds of flesh grow where only one grew before, should be the aim of the farmer, the philanthropist, and the patriot. In the attainment of this object no breeder of past or present times has been so successful as Mr. Booth of Warlaby. Indeed, he had many animals with bosoms so massive, that, as the able writer of 'The Herds of Great Britain,' has well expressed it, "they looked as if they required another pair of fore-legs to support them." And I think it may be shown that utility and beauty are not disconnected in this case; that the conformation of neck objected to in the Booth cattle not only betokens superior constitution and aptitude to lay on flesh on every part of the frame, but is also the most symmetrical.

Experienced graziers are well aware that the light and elegant neck, so much lauded by some, and indeed so desirable in breeds whose sole merit consists in their dairy properties, is usually accompanied by general lightness of flesh, and, in animals in moderate condition, almost invariably with rough shoulders, prominent in their points and bare of flesh. Now it will be admitted that angular points are inconsistent with the conditions of true beauty, which require that the parts should melt gradually and insensibly one into another without any projections.

These conditions a Booth animal generally fulfils. The neck increases rapidly, though not abruptly in size, until it melts insensibly into the shoulders and wide projecting brisket, which again blend imperceptibly with the crop, fore-flank, and ribs, without any depressions or protuberances. When the animal walks, the elbow joint is scarcely if at all seen, and there is no hollow behind it. The motion of the shoulder blades and shoulder points is imperceptible, the former being laid snugly back into the crops, the latter hidden by the full neck vein, which blends, as I have said, with the muscles of the shoulder, neck, and brisket, forming gently tapering lines to the head and breast end. And, indeed, this perfection of fore-quarter in the Booth cows, so far from being restricted to those which are most highly fed, appertains to every individual, and is a distinguishing characteristic of the Warlaby herd.

Mr. Booth attached much importance to the heads of his animals. Conforming, as regards cows, to the popular opinion that they should be moderately small and tapering, he contended that the bull should not only be broad across the brows, but adorned with a "lusty horn," especially stout at the base. Mr. Booth would not use a bull in which these substantial evidences, as he regarded them, of vigorous constitution and procreative power, were wanting. And, indeed, sound physiology teaches that the more or less luxuriant growth of the horn is the result of constitutional operations. The marked influence of ill health or castration on the growth of horn is sufficient proof of this. That the use of sires exhibiting these indications of masculine character has no influence on the female progeny prejudicial to their feminine mien and character, a glance at the Warlaby cows and heifers will show. They are remarkable for their lady-like aspect, and graceful, well-curved, waxy horns, those inextinguishable tell-tales of some otherwise unsuspected jump in the dark, and *of* the dark—inky horns and dingy noses—being unknown amongst any of the leading families of Warlaby. The mild prominent eye is expressive of that equable contented temperament so favourable to the attainment of ripe condition; a tendency further indicated by the double-chin-like appendage of pendulous fat beneath the root of the tongue, which, however objected to by some admirers of the more horse-like conformation of head, gives, in the opinion of others, an engaging piquancy of expression, and is always regarded approvingly by the knowing grazier as an earnest of aptitude for kindly feeding.

Having discussed the salient points of the Warlaby cattle in detail, it may be added that the *tout ensemble* comes perhaps nearer the established standard of perfection than that of any other tribe of Shorthorns.

Although we do not expect a cow to be made by geometrical proportions, yet there are undoubtedly certain established formalities of structure which an experienced judge of shorthorns will pronounce to be indispensable. It has been laid down by a great authority north of the Tweed, that the nearer a Shorthorn approaches to the figure of a parallelopiped, the nearer it is to perfection. Now, this is a geometric solid contained under six parallelograms or planes, of which the opposite ones are similar and parallel. Thus the level plane of the back should find its counterpart in the almost equally broad and level plane of the "under belly"; the two sides should be parallel, or as nearly as possible equi-distant in every point, while the square presented behind by the thighs and twist should match with a corresponding squareness of frame in front.

In conformity with this doctrine, it is evident that the

upper, middle, and lower parts of the shoulders, ribs, and thighs should be in a line, so that if you lay a wand along them it should touch in every point. Apply this test to any fair specimen of a Warlaby cow, and it will be found to stand the trial: even behind the elbow, or hockster joint, there is no hollow space; all is filled up with the heavy flesh, which sheep-breeders call "fore flank," and butchers term "the plates."

If Mr. Booth's preference, like Mr. Douglas's, was for moderate-sized cows, still more decidedly did he advocate the use of moderate-sized bulls. A great bull was, in his opinion, a great evil—not only because large animals are generally wanting in that compactness and symmetry of form so essential in a sire, but because the size of the offspring being generally proportioned to that of the male parent, one of three evils frequently results —either the fœtus fails to receive adequate nourishment from the dam for the support and increase of its growth and strength; or, through lack of sufficient expansive power in the uterus, if not prematurely expelled, is warped in its proportions; or, from its unsuitableness in size to the organs of the dam, the calving is attended with difficulty and danger, and the ligaments of the pelvic cavity, through which the calf must pass, are frequently lacerated, to the permanent injury of the dam.

In an admirable article, which in January, 1863, appeared in the *Mark Lane Express*, 'On the Breeding of Hunters and Hacks,' occurred the following extracts from 'Cline on the Form of Animals,' bearing on this subject: "It has generally been supposed that the breed of animals is improved by the largest males. This opinion has done considerable mischief, and would have done more injury if it had not been counteracted by the desire of selecting animals of the best form and proportions, which are rarely to be met with in those of the largest size. Experience has proved that crossing has only succeeded in an eminent degree in those instances in which the females were larger in the usual proportion of females to males, and that it has generally failed when the males were disproportionately large. When the male is much larger than the female, the offspring is generally of an imperfect form."

It has not unfrequently been alleged that the Booth animals are liable to sterility. If this charge is made against the breeding herd of Warlaby, it would be easy to show, and, indeed, has been shown, that no cattle in the kingdom are more remarkable for their early, regular, and long-continued fecundity; but if it be intended to apply to animals which have attained to a precocious degree of fatness before being put to the bull, or to animals necessarily pampered to meet the absurd but imperious exigencies of the show-field, the remark is, to some extent, true, and equally true of every other herd where the same causes are in operation. If we hear less of the unprolificness of other herds, it is simply because they are of less notoriety and less value, and the subjects of it are, in general, quietly consigned to the butcher, instead of being allowed, as in the case of Mr. Booth's cattle, to continue to live on for years, to adorn the pastures and gratify and astonish the eye of the visitor. Those who have seen amongst the breeding herd at Warlaby such prize animals as Charity, Queen Mab, Nectarine

Blossom, Bride Elect, Lady Grace, Queen of the Ocean, Queen of the Vale, and Soldier's Bride, might not unfairly argue that Mr. Booth's shorthorns must have inherited an unusually healthy organization to be able to resist the baneful effects of that show-yard condition, which certainly impairs the fruitfulness of the majority of animals on the prize-lists. There can be no question but that a redundancy of flesh is unfavourable to fertility, and, when prematurely or artificially acquired, frequently results in barrenness or abortion; but not more frequently in the Warlaby herd than would be the case in any other in which the same propensity to its rapid acquisition existed. It is, however, this propensity which, if it sometimes impairs the usefulness of the Warlaby, as of other *female* animals, enhances the value of the bulls, and especially fits them for their destined purpose—the crossing with other herds. For in the great majority of Shorthorn herds, it may safely be averred, there is a deficiency of that fattening capacity and substance which the Warlaby bulls are so well qualified to impart. There cannot be a more striking illustration of the degree in which the Warlaby bulls imprint on their progeny their own character and qualities than is afforded by the bullocks which are yearly bred and grazed at Warlaby for the Christmas markets. These oxen (though generally the offspring, by Warlaby bulls, of nurse cows selected solely for their milking qualifications, without any regard to form) are fine level animals of remarkable symmetry and substance, and superior quality of flesh, and possess extraordinary facility in acquiring it. A lot of six, calved in 1862, were never under cover after the first winter. Their keep had been straw and rough hay with turnips in winter, and good pasture in summer, with a small allowance of oilcake during the autumn, when the aftermath was finished. They averaged, when slaughtered, upwards of 80 stones of 14 lbs. They were chiefly by Sir Samuel, a closely in-bred bull, being by Crown Prince, and out of Crown Prince's dam, and bore, in a remarkable degree, the Warlaby character—another instance of the use of a bull so bred resulting in the reproduction in the offspring of the family characteristics of the sire.

This leads me to say a few words on a subject on which Mr. Booth had many prejudices to contend with. There are not a few who, disregarding the evidence of facts that no cattle are more robustly organized than those of Warlaby, have jumped to the conclusion that they must be bred from too close consanguinities. It is but fair to hear what was Mr. Booth's own opinion on this subject, and his reasons for it. Some time ago, accepting the then prevalent notion that the Warlaby herd would derive benefit from fresh blood, I ventured to suggest to Mr. Booth the expediency of adopting a cross, when I was met by arguments to the following effect: "Have any of the evils which are usually attributed to in-and-in breeding manifested themselves in my herd? Is there any degeneracy in size, substance, or vigour in the animals? any tendency to premature age? any lack of milking or thriving disposition in the cows, or of capacity of frame, or hardiness of constitution? Are the bulls (with one or two exceptions, such as must occur in every herd) deficient, in masculine character and efficacy, or the sires of a puny or feeble offspring? or do they early become worn-out and unfruitful? And lastly, is there any tribe of Shorthorns that attains a higher or even the same degree of condition on the same food?" All of which questions it being impossible to answer otherwise than in the negative, Mr. Booth added, "Because, if not, it is clear that the only consideration that would justify me in having recourse to a cross, would be the discovery of a tribe which, besides possessing in an equal degree with my own the qualities we have mentioned, are superior to them in utility and symmetry." "Granted," I replied. "Where is it?" was the pertinent but perplexing rejoinder. "The result of the last three crosses upon which I ventured," continued Mr. Booth—"Water King, Exquisite, and Lord Stanley—has made me distrust the policy of any further step in that direction; nor have the results I have witnessed of the experiments of others in crossing animals of my blood with the most fashionable bulls of other strains tended in any instance to remove that distrust."

Any one who considers what careful judgment and patient perseverance are required to ingraft upon a herd of cattle any desired modifications of form and character, and who knows how difficult it is to maintain these acquired characteristics with anything like uniformity (even when breeding exclusively from animals which have themselves inherited them), must see what risk there is of losing them altogether by crossing with extraneous blood, unless such crossing is pursued to that moderate extent only, and with that consummate judgment which the Booths displayed in infusing into their herds the blood of Mussulman and Lord Lieutenant. But Mr. Booth was of opinion that, even if this were not the case, and the infusion of fresh blood were to result in increased prolificness to his own herd, it would materially lessen the value of his bulls for crossing other herds. That value he held to consist in the fact that they have been bred from parents with similar qualities and tendencies for successive generations, and that thus their individual and family characteristics have become so far intensified as greatly to enhance their impressive power. The great axiom of breeding, which nearly 2000 years ago was thus poetically rendered—

> " Fortes creantur fortibus et bonis,
> Est in *juvencis*, est in equis patrum
> Virtus "—

Or, in other words, the principle that " like begets like," can only be relied upon as a governing law where the excellences of the sire are not accidentally but hereditarily, and therefore constitutionally, his. So long as the Warlaby herd retains its present high degree of perfection, together with its hardiness and soundness of constitution, it is certainly desirable to be extremely cautious of diminishing this hereditary transmissive influence. Few, indeed, who have experienced the value of the Warlaby bulls can doubt that Mr. Booth exercised a wise discretion in refusing to expose to such mischances as an incongruous cross might entail the future of a herd whose benefits have been recognised throughout this kingdom, in her remotest colonies, and in many a foreign clime. The hereditary character and virtues of the Warlaby strain are as yet unimpaired—

> " The fortune of the family remains,
> Though grandsires' grandsires the long list contains."

May it long continue to flourish, lending as heretofore, its symmetry and its substance to numberless herds in this and in other lands; and may future generations of the Booth family emulate their sires and their grandsires in the noble work in which they have excelled for three-quarters of a century—the work of exercising their talent and their skill for the benefit of the human race!

In pursuing this important subject I have been called upon to travel over, however briefly, the whole period of Shorthorn history, from the first faint dawnings of improvement down to the present time; for during the whole of this period the Killerby, Studley, and Warlaby herds of Booth cattle have presented a striking mark. Casual allusions have been made to the different fluctuations in price to which, during this time, the shorthorn market has been subjected; and, perchance, a few remarks on its probable future prospects may not be out of place here.

Many years since, a nobleman who was distinguished as being at the same time one of the most far-seeing of statesmen and one of the most eminent and scientific of breeders—the late Earl Spencer—declared that shorthorn-breeding was then only in its infancy, and that an enormous field at home, in the colonies, and in the world at large, would, in course of time, open to reward the science and the energy of English breeders. To a considerable extent this prediction has since been realized; but there are not wanting abundant indications to show that the future is pregnant with yet more abundant promise. The increased intercourse of nations, and the desire in all lands for new and improved breeds of cattle are everywhere most marked, and yet on the whole are only just commencing; while, so far as we can see, though in certain localities other breeds may be partially useful, the English Shorthorn is the only one that can supply the universal want. Whole continents demand them; and though for a time wars and rumours of wars, and a destructive murrain, have prevented their wishes being carried out, the return of peace and disappearance of the cattle plague can but increase the necessities of the nations. Even at home a large and ever-extending field is yet open to the shorthorn breeder. These cattle are rapidly becoming the national breed of Ireland; Scotland finds it her interest to use them most extensively for crossing with her native breeds; and in numerous English counties, where the mongrel and the ne'er-do-weel have hitherto prevailed, the farmers are beginning to understand that their unthrifty and ungainly cattle are not adapted to new systems of agriculture and improved modes of feeding, and are opening their purse-strings for the purchase of the shorthorn. Nothing can prove this more clearly than the enormous prices realized at recent sales of Shorthorn cattle, particularly Mr. Betts's and Mr. MacIntosh's, and that not alone by animals of one kind of blood—Colonel Towneley's Roan Duchesses, and Sir Charles Knightley's Rosys almost vieing with the Grand Duchesses and the Oxfords. Surely, our prospects are most encouraging, and I am entitled to say to my shorthorn friends, "Go on, and prosper."

In concluding these remarks upon the Booth cattle, and upon the subject of breeding so inseparably connected with them, I am fully sensible of the imperfections which must necessarily cleave to a work partly treating of past times, the memory of which is rapidly vanishing, and partly of the present times, in which rivalry, competition, and difference of opinion inevitably prevail. I must rely therefore upon the indulgence of the public, which is well acquainted with the difficulties of the undertaking. I have studiously confined myself, as the subject required, to the Booth cattle and their descendants; and if the herds of other breeders have not been mentioned, which they could not be appropriately, neither have they been depreciated. This merit at least I may claim—to have extenuated nothing, nor " set down aught in malice." And if I should be thought to have succeeded in adding a page to shorthorn lore, to have contributed a trifle to the encouragement of shorthorn breeders, or to have in any degree directed the attention of the public to a subject in which it is so deeply interested, I shall be abundantly rewarded. And so I commend my labours to the indulgence of the reader.

APPENDIX.

[Abridged from "Shorthorn Intelligence," in Bell's Weekly Messenger of April 29, 1867.]

NEARLY ten years have passed since our first visit to Warlaby. That day, as all such days are to lovers of shorthorns, is a day to be remembered. Notwithstanding incessant and tremendous rain, so disastrous generally to the bloom and good looks of cattle, the shorthorns of Warlaby seemed hardly disparaged by the heavy downfall. To worse shapes than *theirs* it would have been absolutely fatal. One after another, they were each seen, approached, examined—in all their prominent characteristics, and in all their minute details. Bianca was there—Bride Elect, in her prime—Royal Bride, her daughter—Satin, old but sprightly, lengthy and well-framed, her long horns fantastically wreathed aloft; her udder, beautifully shaped, and equal to the demands of two calves—Sarcenet, her offspring, by Crown Prince, large and grand, but somewhat loose in structure—Charity, though well advanced in years, displaying the compactness of her youth, and the fulness of her youth's loveliness—Queen of the May, rather past the point of perfection, but still beautiful, and still as round as a cylinder, "as round as a hoop" Mr. Booth used to say—Queen Mab, her career of triumph near at hand—Bridesmaid, with her sweet little head, her sloping shoulders, her wonderful back, and her short legs—Blithe, remarkable for arched ribs and general neatness of shape—Princess Mary and Blithesome, mighty fabrics of symmetry and flesh—Orange Blossom, an astonishing animal for bulk and beauty—Own Sister to Windsor, a great favourite of Mr. Booth's—Red Rose, at one time not less noticeable for personal charms than for fertility at all times—Vivandiere, Campfollower, Lady Blithe (then a calf), and Lady Grace and Queen of the Isles, yearlings. How shall the various and respective peculiarities of these individuals be distinguished? They were all different, yet all belonged to one type; and again and again, even as a man turns to passages in a great writer which arrested attention on a first perusal, they were turned to with renewed delight, for purposes not of delight merely, but of study and instruction. Well; what have the ten years which came and went since first we knew the herd at Warlaby accomplished? What have they done, either to improve it, or worsen it, or leave it *in statu quo?* A recent visit enables us, so far as our own judgment is concerned, to answer the question with some degree of positiveness. There are not so many *astonishing* cows and heifers as there were then; and this, inasmuch as it may be due in no slight measure to the rinderpest-interval of non-forcing, is so far an advantage; but there *are* cows and heifers of *the former stamp exactly;* and the general material of the herd we have no hesitation in pronouncing as good as ever. Whether the *specimen-cows,* as those which are comparatively "made-up" may be termed, or the unforced breeding stock, are considered, we can see no indications of personal degeneracy; and the great number of calves and of cows in milk would seem to afford credible evidence of the general fertility of the herd. The prime old favourites are gone; but their places are taken by younger ones, and are filled with examples of the hereditary type; examples that will be, ten years hence, what the best of those we have mentioned were, ten years ago. One thing struck us, and it will doubtless strike others under similar circumstances. The female lines which *used* to be regarded as the best, no longer take the lead. Instead of the *majority* of the choicest animals being derived from the Isabella, Strawberry and Halnaby, and Blossom families, they are now of the Bliss or Broughton tribe, and a great many very good ones descend from the Christon sort. Mr. Booth attributes to the show-system the decline or the loss of the once prolific kinds. Unquestionably that system is responsible for much evil. Captain Gunter has wisely determined to "show" no more; and we sincerely hope, in the interests of shorthorn breeding, that Mr. Booth may form the same resolution. Neither of the herds, either that at Warlaby or that at Wetherby, requires so great a sacrifice as continual "training" implies; for the reputation of both is as high as it can be. Southey, in one of his delightful letters, says to Sir Walter Scott, "We have both got to the top of the hill, by different paths, and meet there not as rivals but as friends, each rejoicing in the success of the other." No one disputes their place; nor would fifty more rosettes and cups, and fifty more ruined constitutions, give to their position one tittle more of glory.

* * * We propose, contrary to our usual practice, to notice some of the bulls before we request the reader's attention to the cows. The oldest bull at Warlaby, Lord of the Hills (18267), a son of Sir Samuel and Red Rose by Harbinger, (through the sires, we may observe, Mr. Booth retains the blood of his extinct female races), is about the best we ever saw there; *one* of the best, certainly. With plenty of size (he is, in fact, rather large than otherwise), he is wonderfully good in the shoulders, and level along the top; his loins being strong, thick, and broad. Through the heart, behind the scapula, and at the fore-flank, he has all the old Warlaby characteristics, and the under-line is as straight as a wand. His head, neck, and breast, are masculine, and shew breed; his hind-quarters neat; his legs straight, and widely separated. In colour he is roan, of a grizzly-grey sort. His son, Mountain Chief (from Soldier's Bride), a great tall roomy bull of a rather deeper roan colour, cannot be described as *handsome*, nor yet as anything but a *good* animal. He has the Booth well-rounded fore-rib, a firm level back, and abundant depth of fore-quarter. Commander-in-Chief (21451), roan, the son of Valasco and of Campfollower by Crown Prince, just now in full perfection of symmetry and muscular development, is a magnificent representative of the genuine Warlaby type of shorthorn. Imposing in presence, he comes out with a lordly air of dignity and self-assertion; quiet, unpretending, self-sustained. In form, he is of the thick substantial order; and if any points demand special notice, the crops do, for their wonderful width. Among other remarkable features of detail, the filling-up between the crops and the shoulder-blades is wonderfully complete and excellent; his loins are of astonishing thickness and strength; the breast has immense development; and the flank is of the most extraordinary weight we ever felt. The thighs are wide and heavy, connected by an ample twist; and the arms, of vast power, contrast with the fineness of the bone below the knee. The bull's way of walking deserves to be mentioned; fore and hind legs are placed so grandly apart, are so rightly made, and move forward in such a direct line when he steps. On our observing this to Mr. Booth, his reply was highly descriptive; " Yes; he walks with his legs well *outside* him." With The Sutler (23061), who served a year at Mr. Bruere's, we were " agreeably disappointed," having heard him disparaged, and in no very qualified terms. In consequence of his teeth being loose and out of order, he did badly for some months, but they are right now, and he is rapidly growing a fine bull. The rest of the bulls are chiefly calves; very good, but too young perhaps for particular and discriminative delineation. Yet a few months will make them worth from a hundred to two hundred guineas a year, and, what is more, will find them earning it. One of them, only ten months old, from Lord of the Valley and Charlotte, exacts special mention and special praise. As large as most yearlings, he is very handsome, a full rich roan, true in build, and stylish. Great Hope, a red, from Hopeful, and by Commander-in-Chief, is also an admirable calf, destined, we expect, to redeem the promise of his lineage and present appearance. * * * Three magnificent cows were out upon the long middenstead opposite the line of boxes with which all visitors to Warlaby are familiar. They were Queen of the Ocean (red and white), Lady Joyful (roan), and Soldier's Bride (white); and they presented, not only an agreeable contrast in colour (one of the acknowledged attractions of shorthorns), but also a fine study as respects excellence of form exhibited under three distinct varieties. They represent also, three distinct female lines; that of Strawberry and Halnaby; that of Bliss or Broughton; and that of Vivandiere or the Moss Roses. Queen of the Ocean, the First Prize cow at the Battersea Royal, retains much of the loveliness of her show-yard days. Lower in condition than when we saw her at the North Lancashire with Pride of Southwick, another Royal Prima Donna, she has not given way or become censurable anywhere. The only difference seems to be that she has rather less flesh all over; equally disposed it is, but there is less in quantity. She shows herself as well as a cow can do, and looks faultlessly beautiful. Unfortunately, not breeding, nor likely to breed, for we suppose it is no scandal to say so, she is little more than a picture now; but the power of her blood imparted by her son, Prince of Battersea, to Blooming Bride, a noble heifer, may be expected to work with its hereditary tendency towards the truest ideals, and contribute to the reproduction of animals resembling herself. Soldier's Bride, the least handsome of the three, is still a noble cow; of wonderful length but with unyielding firmness of back. Her failing point is of course, as those who knew her in her palmy days will suspect, the droop of the hind-quarters; and this, at her age, is not likely to improve. She is very heavily fleshed. The last of the triad, Lady Joyful, is a truly splendid cow, uniting enormous weight with elegance of shape. Wide and deep, displaying the best Booth points throughout, she has a style about her which even sooner than her structural merit commands admiration. This style is explained upon a piecemeal examination. Her head is beautiful, and the shape and setting-on of her neck, and the moulding of her bosom, neck-vein, shoulders, and chine, are simply examples of perfection. Her dam, Lady Blithe, has bred perhaps more really first-rate shorthorns than any other cow in existence. She has replenished the Warlaby herd, and maintained its superiority by supplying some of the best females the herd now contains. To Lord of the Valley she has bred eight calves, and to Sir Samuel she bred once; nine calves, and her tenth year will not be completed till July. The alliance of Lord of the Valley with Windsor's stock appears to have answered admirably in every instance. In the produce of Lady Blithe this is very noticeable, and also in Bride of the Vale, the daughter of Soldier's Bride. Lady Blithe is a thorough high-bred dairy cow; happily (for the fact exempts her from many risks) she has no *showyard* pretensions whatever; but a well-knit frame, and a look of pure and ancient blood about the head and horns, and a neck of peculiarly graceful and feminine outline (a point of great importance), distinguish her at once as a shorthorn of note. Her daughter, Lady Mirth by Sir Samuel, a roan seven-year-old, has the tidy cow-like character of her grandam Blithe by Hopewell, and is the mother of several members of the herd, one of which, a very gay-looking red heifer, Lady Jane by Lord of the Valley, specially won our regard. Merry Peal, Lady Jane's own sister, and Merry Monarch (22349), are among the offspring of Lady Mirth. Lady Jane (short, we think, of two years old) reminded us forcibly of Mr. Carr's Lady of the Valley when a yearling, closely resembling her about the head and fore-quarters; but Lady of the Valley, we think, was a better heifer than her half sister, though an uncommonly good one, is at present. Three more of the valuable Blithe family (and contracting space forbids the mention of more than three) must be particularized; Lady Fragrant (the Royal prize-taker), and the twins Lady Grateful and Lady Gratitude, own sisters to Lady Fragrant and Lady Joyful. The first of these three, and the greatest both in fame and personal loveliness, has changed but little since her triumphant appearance at Plymouth. Though not large, she is exquisitely beautiful; finely rounded along the saddle to the shoulders, with a bright countenance and a sweetly formed neck. It would be difficult to imagine symmetry more perfect. She has bred a lovely red heifer calf, Lady Perfume, to Commander-in-Chief. Lady Gratitude, now in calf to Mountain Chief, has the red roan colour so common in the Blithe family; capital flesh, good hair, and widely sprung ribs. Lady Grateful is a still better heifer; all covered with rich roan; has loins of extraordinary excellence; a back admirable throughout; flank, fore-flank, and all the under-line, beautiful. She gave birth a few weeks ago to a wonderfully fine roan bull calf, which, if luck attend him, will by and by stand foremost in the first class of Warlaby sires. Of the Christon family, Alfreda by Prince Alfred, a good cow of the deeply fleshed old Booth sort, and the dam of Prince Christian (22581), now "out" at Mr. Aylmer's—Mabel, a handsome roan yearling by General Hopewell; and Melissa, a white three-year-old by the same bull—are our favourites; and, in sub-selecting, Mabel would be chosen. She is neat, compact, and cow-like. The once famous Halnaby tribe is represented by Modesty 2nd, a smart, stylish, light-boned cow of 1863, by Lord of the Valley and from old Modesty—and by Prudence, a splendid white yearling heifer from her by General Hopewell; and the prestige of the Bianca branch of the family depends upon two fine animals, Royal Bridesmaid by Prince Alfred, and Blooming Bride, before alluded to, a handsomely formed and thorough Warlaby heifer by Queen of the Ocean's short-lived son, himself a winner, Prince of Battersea. If the Vivandiere or Moss Rose tribe, to which we now turn, had only Bride of the Vale (the daughter of Soldier's Bride) to represent it, the acknowledged value of the family would be amply asserted. Bride of the Vale, a rich roan, by Lord of the Valley, and of great size, is one of the best heifers at Warlaby; and though only two years and three months old, bred a calf about the middle of March. She has, proportioned to the difference in age, the length of her dam, and all her dam's grand properties, with greater evenness; and whether regarded as fitted to be a show animal with the highest expectations, or considered with reference to the more profitable object of multiplying her kind, she is indeed a superb heifer. With very little forcing she would prove equal to nearly any of the winners that Warlaby has sent out. She has both style and excellence of detail. It is worthy of note, that about Lord of the Valley's stock there is a brightness of aspect and a loftiness of carriage which infallibly add to the attractiveness of the massive and substantial Warlaby cows. In Soldier's Daughter, the offspring of him and Campfollower, these characteristics are strikingly apparent; but we mention her, not so much on account of her merit, which is however of a distinguished order, as because it gives us an opportunity of stating, in reply to reiterated assertions to the contrary, that not only is the tribe to which she belongs, but that all the tribes composing the Warlaby herd, are famous for possessing more than ordinary milking powers. Campfollower was an extreme milker, and, indeed, died of milk fever after giving birth to Soldier's Daughter. * * * One other family, consisting of only three members, deserves especial mention, for its history is interesting. We allude to the cows British Rose by Prince George; her daughter, Wild Rose by Lord of the Valley; and a calf, Christmas Rose, Wild Rose's sister. These animals are derived, the former through four, the second and third through five crosses, from a red polled Galloway Scot; and the experiment, issuing as it does in pure shorthorn characteristics, is entirely successful. * * * Such, in conclusion, are our thoughts upon the herd at Warlaby. It is a magnificent herd; a herd containing animals of the most exalted merit, and representing family alliances and combinations of blood, long and watchfully proved, which guarantee, under circumstances of ordinary fortune, a continuance of animals of the same stamp. It is the result, not of capricious and versatile tastes, but of definite principles thoroughly examined and deliberately adopted; and it has the general uniformity which proceeds from such causes. But general uniformity is consistent with gradations of excellence, and it is so at Warlaby. The term "good," like the term "bad," has a worst and a best. A friend of ours—not now a shorthorn breeder—with no class-prejudices, with no class-partialities, but a lover of fine shorthorns whatever distinctive name they bear, intimately acquainted with the leading herds of the day, and competent to form a sound judgment respecting the characteristic merits of the species and the particular merits of individuals, strongly contends that he could select from Warlaby at least half a dozen females, not only better than any other single herd could supply, but better than could be selected from all the herds of Great Britain and Ireland in the aggregate. We give this emphatic opinion without comment.

REMARKS ON THE BREEDING OF FAVOURITE, COMET, AND THE EARLIER BOOTH BULLS. By the Rev. J. STORER.

Few people have any idea of the amazing extent to which in-and-in-breeding was carried on by the Brothers Colling; and so great was the complication it involved, that few of those who know the outline of the circumstances, can adequately realize all their intricacies. It is almost impossible to describe even proximately in some of its stronger features the system they pursued. But the attempt ought to made; for the Messrs. Colling's system of in-and-in-breeding, is not only one of the most remarkable and authentic cases in the history of the reproduction of animals with which we are acquainted, but the earlier Booth Bulls were amongst those most strongly subjected to its influence.

Mr. C. Colling's bull Bolingbroke, and his cow Phœnix, were brother and sister on the sire's side, and nearly so on the dam's. They were of the same family; and the only difference in descent was, that Bolingbroke was a grandson of Dalton Duke, while Phœnix was not. But this apparent difference, slight as it is, was not all real; for Dalton Duke also contained some portion of their common blood. Arithmetically stated, the blood of the two being taken and divided into *thirty-two* parts, *twenty-nine* of those parts were of blood common to both, rather differently proportioned between them. Phœnix had *sixteen* of those parts, Bolingbroke *thirteen;* the latter having also *three* fresh parts derived from Dalton Duke, which made up the *thirty-two*.

Being thus *very nearly* own brother and sister, they were the joint parents of the bull Favourite. That bull was next put to his own mother Phœnix, so nearly related to him on his sire's side also; and the produce was Young Phœnix. To this heifer Favourite was once more put, she being at once his daughter and *more than own sister too*. For their two sires, Bolingbroke and Favourite, were not only as nearly as possible consanguineous with each other, but also with the cow Phœnix, to which they were both put. The result was —Comet.

APPENDIX.

Nor was this all. The system was carried much further. The celebrated Booth Bull Albion was not only a son of the in-and-in Favourite-bred Comet, but his dam was a grandaughter of Favourite on both sides, and descended besides from both the sire and the dam of Favourite.

It is not so possible to make an exact statement with regard to Pilot, for it is not known whether he was by Major (398), or Wellington (680). Nor does it much matter; for five-eighths of Major's and three-quarters of Wellington's blood were derived from Favourite, by repeated inter-crossings; and Pilot's dam was not only by Favourite but she was also the grandaughter of Foljambe, the sire of both the parents of Favourite.

Marshal Beresford was, like Albion, a son of Comet; and his dam was by a grandson of Favourite out of a daughter of Favourite.

Suworrow was by a son of Favourite; and his dam was a daughter of Favourite; and Twin Brother of Ben was from a cow by Foljambe, the double grandsire of Favourite.

Even this does not exhaust the subject. Many of the above-mentioned animals were otherwise related to each other by a common descent from Hubback, and from other progenitors.

Albion has been called "The Alloy Bull." I think with very little reason. When it is remembered that he is the *seventh* in descent from that blood, and that therefore only *one* part of his blood came from "The Alloy," against *one hundred and twenty-seven* parts which were not derived from it, the chances against either good or evil resulting therefrom were infinitesimally small; and so no doubt such an acute observer as Mr. Booth well knew.

The following is an imperfect List of the Prizes won by the WARLABY AND KILLERBY SHORTHORNS, *no record having been kept of the earlier Prizes, nor until very recently of those taken at the minor Shows.*

1840. YORKSHIRE SHOW, AT NORTHALLERTON.

Second prize bull calf, Leonard.
First prize three-year-old cow, Bracelet.
Second prize one-year-old heifer, Mantalini.

1841. R. A. S. AT LIVERPOOL.

First prize cow, Bracelet.
First prize two-year-old, Mantalini.

YORKSHIRE, HULL.

First prize cow, Bracelet.
Second prize two-year-old heifer, Mantalini.

HIGHLAND SOCIETY, BERWICK-ON-TWEED.

First prize cow, Bracelet.
First prize two-year-old heifer, Mantalini.

1842. R. A. S. BRISTOL.

First prize cow, Necklace.

YORKSHIRE, YORK.

Second prize cow, Necklace.
First prize three-year-old, Mantalini.
Extra prize, Bracelet.
Sweepstakes, best cow, Necklace.

Sweepstakes, best lot of four animals :—
  Bracelet, Necklace, Mantalini and Ladythorn.

1843.   YORKSHIRE, DONCASTER.

First prize cow, Necklace.
Second prize three-year-old cow, Faith.
First prize two-year-old, Ladythorn.
Second prize two-year-old, Birthday.
Second prize yearling, Modish.
Extra stock second prize, White Strawberry.

1844.   R. A. S. SOUTHAMPTON.

First prize cow, Birthday.
First prize yearling, Bud.

   YORKSHIRE, RICHMOND.

Second prize cow, Faith.
First prize three-year-old cow, Birthday.
First prize two-year-old, Modish.
Second prize yearling, Bud.
First prize heifer, Pearl.
First and second extra prize, Bracelet and Necklace.

1845.   R. A. S. SHREWSBURY.

First prize, Ladythorn (sold to and shown by Mr. Banks Stanhope).

   YORKSHIRE, BEVERLEY.

Second prize aged bull, Fitz-Leonard.
First prize cow, Birthday.
First prize two-year-old heifer, Hope.
Second prize yearling, Gem.

1846.   R. A. S. NEWCASTLE.

First prize cow, Hope.
First prize two-year-old, Gem.
Extra prize, first, Necklace.
Extra prize, second, Birthday.

YORKSHIRE, WAKEFIELD.

First prize yearling bull, Hamlet.
First prize cow, Mantalini.
Second prize cow, Alba.
First prize three-year-old cow, Hope.
First prize two-year-old, Gem.
First prize calf, Bloom.
First prize extra, Birthday.

SMITHFIELD CLUB.

First prize cow, Gold Medal for best female in yard, Silver Medal to breeder, Necklace.

1847. ### R. A. S. NORTHAMPTON.

First prize cow, Cherry Blossom.
First prize two-year-old heifer, Isabella Buckingham.
First prize yearling heifer, Charity.
Second prize bull, calved previous to 1845, Hamlet.

YORKSHIRE, SCARBOROUGH.

First prize cow, Hope.
First prize three-year-old cow, Cherry Blossom.
First prize two-year-old heifer, Isabella Buckingham.
First prize yearling, Charity.
Second prize extra stock, Bagatelle.

1848. ### R. A. S. YORK, AND THE LOCAL IN THE ALL ENGLAND CLASS.

Second prize cow, Isabella Buckingham.
First prize two-year-old heifer, Charity.
First prize cow (Local prize), Hope.
First prize pair of heifers, Charity and British Queen.

1849. ### R. A. S. NORWICH.

First prize cow, Charity.
Second prize Isabella Buckingham.

YORKSHIRE, LEEDS.

First prize yearling bull, Hopewell.
Second prize bull calf, Bullion.
First prize cow, Isabella Buckingham.
Second prize cow, Bagatelle.
First prize three-year-old, Charity.
First prize extra stock, Cherry Blossom.

HIGHLAND SOCIETY, GLASGOW.

First prize, Charity.
Second prize, Isabella Buckingham.

1850.   ### R. A. S. EXETER.

First prize yearling bull, Harbinger.
First prize cow, Isabella Buckingham.
Second prize cow, Bagatelle.

YORKSHIRE, THIRSK.

First prize cow, Charity.
Second prize heifer calf, Bride.
Second prize extra stock, Isabella Buckingham.

1851.   ### R. A. S. WINDSOR.

First prize cow, Plum Blossom.

YORKSHIRE, BURLINGTON.

Second prize bull calf, British Boy.
First prize cow, Cherry Blossom.
Second prize cow, Plum Blossom.

1852.   ### R. A. S. LEWES.

First prize two-year-old bull, Red Knight.
First prize two-year-old heifer, Rose Blossom.
Second prize yearling, Bridesmaid.

YORKSHIRE, SHEFFIELD.

First prize bull calf, Windsor.
Second prize cow, Rose Blossom.

First prize two-year-old, Venus Victrix.
Second prize two-year-old, Bride.
Second prize one-year-old, Bridesmaid.

1853.   R. A. S. GLOUCESTER.

First prize yearling bull, Windsor.
Second prize cow, Rose Blossom.
First prize two-year-old, Bridesmaid.
Second prize two-year-old, Peach Blossom.

YORKSHIRE, YORK.

First prize yearling bull, Windsor.
Second prize cow, Rose Blossom.
First prize three-year-old cow, Venus Victrix.
First prize two-year-old, Bridesmaid.

NORTH LANCASHIRE, BLACKBURN.

First prize yearling bull, Windsor—Silver Medal as best male animal, and Silver Cup offered by Col. Towneley.
Second prize cow, Rose Blossom.
First prize two-year-old heifer, Bridesmaid, and Silver Medal as best female animal, and Silver Cup offered by Col. Towneley.

1854.   R. A. S. LINCOLN.

Second prize bull, Windsor.
Second prize cow, Venus Victrix.

Did not show at the Yorkshire at Ripon.

ROYAL IRISH, ARMAGH.

First prize aged bull, and Gold Medal for best bull and Silver Medal to breeder, Windsor.
First prize cow, Bridesmaid.

HIGHLAND, BERWICK-ON-TWEED.

First prize aged bull, Windsor.
First prize bull calf, Prince Alfred.

First prize cow, Bridesmaid.
First prize yearling heifer, Bride Elect.

NORTHUMBERLAND COUNTY, MORPETH.
First prize cow, Venus Victrix.

1855.   R. A. S. CARLISLE.

First prize bull, Windsor.
First prize cow, Bridesmaid.
Second prize two-year-old heifer, Bride Elect.

YORKSHIRE, MALTON.
First prize aged bull, Windsor.
First prize cow, Bridesmaid.
Second prize cow, Venus Victrix.
Second prize two-year-old, Bride Elect.
Second prize heifer calf, Queen of the May.

1856.   R. A. S. CHELMSFORD.

First prize yearling heifer, Queen of the May.

YORKSHIRE, ROTHERHAM.
Second prize cow, Venus Victrix.
Second prize three-year-old cow, Bride Elect.
First prize heifer, Queen of the May.
First prize fat heifer, Water Nymph.
First prize extra stock, Bridesmaid.

1857.   R. A. S. SALISBURY.

Second prize bull calf, Lord of the Valley.
Second prize two-year-old heifer, Queen of the May.
Second prize yearling, Queen Mab.

YORKSHIRE, YORK.
First prize bull calf, Lord of the Valley.
First prize three-year-old cow, Nectarine Blossom.
First prize two-year-old, Queen of the May.

Second prize yearling, Queen Mab.
First prize heifer calf, Queen of the Isles.

Durham County, Stockton-on-Tees.

Second prize cow, Bride Elect.
First prize two-year-old, Queen of the May, and 100-guinea Challenge Cup for best animal.
First prize yearling, Queen Mab.
First prize calf, Queen of the Isles.

Northumberland, Cornhill.

Second prize yearling, Venus de Medici.

1858.

R. A. S. Chester.

First prize cow, Nectarine Blossom.
First prize yearling heifer, Queen of the Isles.

Yorkshire, Northallerton.

First prize cow, Nectarine Blossom.
Second prize two-year-old heifer, Queen Mab.
First prize yearling, Queen of the Isles.
Second prize heifer calf, Queen of the Vale.
First prize extra stock, Bride Elect.
Special prize, 20-guinea Cup, for best animal in yard, Queen of the Isles.

Durham, Sunderland.

First prize cow, Nectarine Blossom, and 100 gs. Cup, for best animal in the yard.
First prize two-year-old heifer, Queen Mab.
First prize yearling heifer, Queen of the Isles.
Second prize heifer calf, Queen of the Vale.
First prize bull calf, Sir James.

1859.

Royal North Lancashire, Blackburn.

First prize cow, Nectarine Blossom.
Second prize cow, Queen Mab.

APPENDIX.

YORKSHIRE, HULL.

First prize three-year-old cow, Queen Mab.

DURHAM, WEST HARTLEPOOL.

First prize cow, Queen Mab, and 100 gs. Cup, for best animal in yard,—thus winning it three years in succession.
Extra stock, Nectarine Blossom.

1860. ### R. A. S. CANTERBURY.

Second prize cow, Queen Mab.

YORKSHIRE, PONTEFRACT.

First prize cow, Queen Mab.
Second prize two-year-old heifer, Queen of the Vale.
Second prize yearling, Soldier's Bride.
Special prize and Silver Cup, for best shorthorn cow, Queen Mab.

NORTHERN COUNTIES, DARLINGTON.

First prize under three years old, Soldier's Bride, and 100 gs. Gold Cup for best animal in yard.
Second prize under four years old, Queen of the Isles.

YORKSHIRE FAT STOCK SHOW, YORK.

First prize, under three years old, Soldier's Bride, and Silver Cup for best animal in yard.

DURHAM, BISHOP AUCKLAND.

First prize one-year-old heifer, Soldier's Bride.

CLEVELAND SOCIETY, MIDDLESBOROUGH.

First prize cow, Queen Mab.
First prize two-year-old, Queen of the Vale.
First prize yearling, Soldier's Bride.

1861. **R. A. S. and Yorkshire combined, Leeds.**

Second prize, Queen Mab.
Second prize two-year-old, Soldier's Bride.

Durham County, Darlington.

First prize two-year-old, and 100-guinea Cup, for best animal in yard, Soldier's Bride.

Highland Society, Perth.

First prize, Queen of the Vale.
Second prize, Queen Mab.
First prize two year old, Soldier's Bride.

Cleveland, Yarm.

First prize cow, Lady Grace.
Second prize, Queen of the Vale.
First prize two-year-old, Soldier's Bride.

Northumberland, Newcastle-on-Tyne.

First prize cow, Queen Mab.
Second prize, Queen of the Vale.
First prize two-year-old, Soldier's Bride.

1862. **R. A. S. Battersea.**

First prize cow, Queen of the Ocean.
First prize yearling heifer, Queen of the May 2nd.
Gold Medal, for best female in the yard, Queen of the Ocean.

Yorkshire Society, York.

First prize, Queen of the Vale.
First prize three-year-old, Queen of the Ocean.
First prize yearling, Queen of the May 2nd.

Durham, Sedgefield.

First prize and 100-guinea Cup, for best animal in yard, Queen of the Ocean.
Second prize cow, Queen of the Vale.
Second prize one-year-old, Queen of the May 2nd.

YORKSHIRE, GUISBOROUGH.

First prize cow, Queen of the Ocean.
Second prize, Sincerity.
First prize yearling, Queen of the May 2nd.

LANCASHIRE, PRESTON.

Second prize one-year-old, Queen of the May 2nd.

CRAVEN, SKIPTON.

First prize, Soldier's Bride.
Second prize, Queen of the Ocean.
First prize two-year-old, Graceful.
Second prize one-year-old, Queen of the May 2nd.
Silver Cup for the best six animals:—
    Queen of the Vale, Sincerity, Soldier's Bride, Queen of the Ocean, Queen of the May 2nd, and Graceful.

1863.   **R. A. S. WORCESTER.**

Second prize two-year-old heifer, Queen of the May 2nd.
Prize for best pair of cows:—
    Queen of the Ocean and Soldier's Bride.
Prize for best pair of heifers:—
    Lady Graceful and Lady Joyful.

YORKSHIRE, STOCKTON.

First prize bull calf, Prince of Battersea.
First prize cow, Queen of the Ocean.
Second prize, Soldier's Bride.
Second prize two-year-old heifer, Queen of the May 2nd.
First extra prize, Queen of the Vale.

DURHAM, GATESHEAD.

First prize bull calf, Prince of Battersea.
First prize cow, Soldier's Bride.

Second prize two-year-old heifer, Queen of the May 2nd.

NORTHUMBERLAND, HEXHAM.

First prize bull calf, Prince of Battersea.
First prize cow, Queen of the Ocean.
Second prize, Soldier's Bride.
First prize two-year-old, Queen of the May 2nd.

CLEVELAND, REDCAR.

First prize cow, Queen of the Vale.
First prize two-year-old heifer, Lady Joyful.

NORTH LANCASHIRE, LANCASTER.

First prize bull calf, Prince of Battersea.
First prize cow, Queen of the Ocean.
Second prize two-year-old, Queen of the May 2nd.

CRAVEN, SKIPTON.

First prize bull calf, Prince of Battersea.
First prize cow, Queen of the Ocean.
Second prize two-year-old, Queen of the May 2nd.

HALIFAX.

First prize bull calf, Prince of Battersea.
First prize cow, Queen of the Ocean.
Second prize two-year-old, Queen of the May 2nd.

KEIGHLEY.

First prize bull calf, Prince of Battersea.
First prize cow, Queen of the Ocean.
Second prize two-year-old, Queen of the May 2nd.

1864.   R. A. S. NEWCASTLE.

Third prize yearling, Prince of Battersea.
Second prize bull calf, British Crown.

YORKSHIRE, HOWDEN.

First prize aged bull, Prince Alfred.
Second prize, Knight of Windsor.
First prize bull calf, British Crown.
First prize yearling heifer, Lady Fragrant.

RIPON.

Second prize bull, Prince Alfred.
First prize yearling bull calf, Prince of Battersea.

CLEVELAND, REDCAR.

First prize bull, Prince Alfred.
First prize yearling heifer, Lady Fragrant.

SCARBOROUGH.

First prize bull, Prince Alfred.
First prize bull calf, British Crown.
First prize one-year-old heifer, Lady Fragrant.

1865. R. A. S. PLYMOUTH.

Second prize yearling bull, Commander-in-Chief.
First prize two-year-old, Lady Fragrant.

YORKSHIRE, DONCASTER.

Third prize bull calf, Master Hopewell.
First prize Champion Silver Cup, for best animal in yard, Lady Fragrant, two-year-old heifer.
First prize extra stock, Prince Alfred.

DURHAM COUNTY, DURHAM.

First prize bull, Prince Alfred.
First prize yearling, Commander-in-Chief.
First prize bull calf, Prince Christian.
First prize two-year-old heifer, Lady Fragrant, and 100-guinea Cup, for best animal in yard.

Northumberland County, Morpeth.

First prize bull calf, Prince Christian.
First prize two-year-old heifer, Lady Fragrant.

Cleveland, Guisborough.

First prize bull, Prince Alfred.
First prize cow, Queen of the May 2nd.

Northamptonshire, Peterborough.

First prize two-year-old heifer, Lady Fragrant.

THE END.

www.ingramcontent.com/pod-product-compliance
Lightning Source LLC
Chambersburg PA
CBHW082330220526
45470CB00008B/2453